Confronting the Core Curriculum

Considering Change in the Undergraduate Mathematics Major

Conference Proceedings

The conference, "West Point Core Curriculum Conference in Mathematics," was supported by the National Science Foundation (DUE 9450767) through the Division of Undergraduate Education. The comments and opinions in this report are those of the authors, and not necessarily those of the foundation.

ISBN 0-88385-155-5

Library of Congress Catalog Number 97-74332

Printed in the United States of America

Current Printing

10 9 8 7 6 5 4 3 2

Confronting the Core Curriculum

Considering Change in the Undergraduate Mathematics Major

Conference Proceedings

Edited by

John A. Dossey

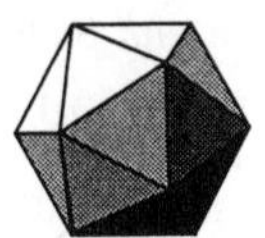

Published by
THE MATHEMATICAL ASSOCIATION OF AMERICA

MAA Notes Series

The MAA Notes Series, started in 1982, addresses a broad range of topics and themes of interest to all who are involved with undergraduate mathematics. The volumes in this series are readable, informative, and useful, and help the mathematical community keep up with developments of importance to mathematics.

MAA Notes

9. Computers and Mathematics: The Use of Computers in Undergraduate Instruction, *Committee on Computers in Mathematics Education, D. A. Smith, G. J. Porter, L. C. Leinbach, and R. H. Wenger,* Editors.
11. Keys to Improved Instruction by Teaching Assistants and Part-Time Instructors, *Committee on Teaching Assistants and Part-Time Instructors, Bettye Anne Case,* Editor.
13. Reshaping College Mathematics, *Committee on the Undergraduate Program in Mathematics, Lynn A. Steen,* Editor.
14. Mathematical Writing, by *Donald E. Knuth, Tracy Larrabee, and Paul M. Roberts.*
15. Discrete Mathematics in the First Two Years, *Anthony Ralston,* Editor.
16. Using Writing to Teach Mathematics, *Andrew Sterrett,* Editor.
17. Priming the Calculus Pump: Innovations and Resources, *Committee on Calculus Reform and the First Two Years,* a subcomittee of the Committee on the Undergraduate Program in Mathematics, *Thomas W. Tucker,* Editor.
18. Models for Undergraduate Research in Mathematics, *Lester Senechal,* Editor.
19. Visualization in Teaching and Learning Mathematics, *Committee on Computers in Mathematics Education, Steve Cunningham and Walter S. Zimmermann,* Editors.
20. The Laboratory Approach to Teaching Calculus, *L. Carl Leinbach et al.,* Editors.
21. Perspectives on Contemporary Statistics, *David C. Hoaglin and David S. Moore,* Editors.
22. Heeding the Call for Change: Suggestions for Curricular Action, *Lynn A. Steen,* Editor.
23. Statistical Abstract of Undergraduate Programs in the Mathematical Sciences and Computer Science in the United States: 1990–91 CBMS Survey, *Donald J. Albers, Don O. Loftsgaarden, Donald C. Rung, and Ann E. Watkins.*
24. Symbolic Computation in Undergraduate Mathematics Education, *Zaven A. Karian,* Editor.
25. The Concept of Function: Aspects of Epistemology and Pedagogy, *Guershon Harel and Ed Dubinsky,* Editors.
26. Statistics for the Twenty-First Century, *Florence and Sheldon Gordon,* Editors.
27. Resources for Calculus Collection, Volume 1: Learning by Discovery: A Lab Manual for Calculus, *Anita E. Solow,* Editor.
28. Resources for Calculus Collection, Volume 2: Calculus Problems for a New Century, *Robert Fraga,* Editor.
29. Resources for Calculus Collection, Volume 3: Applications of Calculus, *Philip Straffin,* Editor.
30. Resources for Calculus Collection, Volume 4: Problems for Student Investigation, *Michael B. Jackson and John R. Ramsay,* Editors.
31. Resources for Calculus Collection, Volume 5: Readings for Calculus, *Underwood Dudley,* Editor.

32. Essays in Humanistic Mathematics, *Alvin White,* Editor.

33. Research Issues in Undergraduate Mathematics Learning: Preliminary Analyses and Results, *James J. Kaput and Ed Dubinsky,* Editors.

34. In Eves' Circles, *Joby Milo Anthony,* Editor.

35. You're the Professor, What Next? Ideas and Resources for Preparing College Teachers, *The Committee on Preparation for College Teaching, Bettye Anne Case,* Editor.

36. Preparing for a New Calculus: Conference Proceedings, *Anita E. Solow,* Editor.

37. A Practical Guide to Cooperative Learning in Collegiate Mathematics, *Nancy L. Hagelgans, Barbara E. Reynolds, SDS, Keith Schwingendorf, Draga Vidakovic, Ed Dubinsky, Mazen Shahin, G. Joseph Wimbish, Jr.*

38. Models That Work: Case Studies in Effective Undergraduate Mathematics Programs, *Alan C. Tucker,* Editor.

39. Calculus: The Dynamics of Change, *CUPM Subcommittee on Calculus Reform and the First Two Years, A. Wayne Roberts,* Editor.

40. Vita Mathematica: Historical Research and Integration with Teaching, *Ronald Calinger,* Editor.

41. Geometry Turned On: Dynamic Software in Learning, Teaching, and Research, *James R. King and Doris Schattschneider,* Editors.

42. Resources for Teaching Linear Algebra, *David Carlson, Charles R. Johnson, David C. Lay, A. Duane Porter, Ann E. Watkins, William Watkins,* Editors.

43. Student Assessment in Calculus: A Report of the NSF Working Group on Assessment in Calculus, *Alan Schoenfeld,* Editor.

44. Readings in Cooperative Learning for Undergraduate Mathematics, *Ed Dubinsky, David Mathews, and Barbara E. Reynolds,* Editors.

45. Confronting the Core Curriculum: Considering Change in the Undergraduate Mathematics Major, *John A. Dossey,* Editor.

46. Women in Mathematics: Scaling the Heights, *Deborah Nolan,* Editor.

47. Exemplary Programs in Introductory College Mathematics: Innovative Programs Using Technology, *Susan Lenker,* Editor.

48. Writing in the Teaching and Learning of Mathematics, *John Meier and Thomas Rishel.*

49. Assessment Practices in Undergraduate Mathematics, *Bonnie Gold,* Editor.

50. Revolutions in Differential Equations: Exploring ODEs with Modern Technology, *Michael J. Kallaher,* Editor.

These volumes can be ordered from:
MAA Service Center
P.O. Box 91112
Washington, DC 20090-1112
800-331-1MAA FAX: 301-206-9789

The present volume would not have been possible without the conditions created by Frank Giordano in the Department of Mathematical Sciences at the U.S. Military Academy during his tenure as Chair of the Department. His leadership, and that of the Academy's Dean and Superintendent provided an environment in which curricular thinking was highly valued and innovations in curricular design a matter of testing, not dreaming. The conference which provided the thoughts contained within this volume was his idea. Don Small served as a co-chair of the efforts to bring that dream to reality. We were backed by the special contributions of Chris Arney, Ron Miller, Walt Seidel, Don Engen, Rich West, Mike DelRosario, Paul Lamakis, Debra Strukel, and Chuck Clark. A grant from the Division of Undergraduate Education at the National Science Foundation provided the financial support. The confluence of these dreams, actions, support, and the individuals participating made the following work a reality.

Introduction

John A. Dossey[1]

The contents of the first two-years of collegiate curriculum for individuals specializing in the mathematical sciences, be it at a community college, a small liberal arts institution, or a large research-oriented university, is a topic of great concern ([2],[6],[7],[13]). For those selecting quantitative-based majors, the primary focus has been on improving the programs for those specializing in mathematics through reforming the calculus, rather than on developing programs that support the full spectrum of undergraduate majors requiring strong quantitative skills. In the past, the rationale offered for the focus on calculus was based on preparation for graduate studies in mathematics or the applications of calculus in science and engineering. Today, the conversation is shifting to meeting the needs of a broad range of college students as they prepare to enter a world of work governed by quantitative models and data displays, run by optimization models, and filled with the need to both comprehend and visualize vast amounts of technical knowledge. The signal to mathematics departments of this change in the world of collegiate mathematics was sounded by Alan Tucker and other members of the Mathematical Association of America's (MAA) Committee on the Undergraduate Program in Mathematics (CUPM) in 1981 with the publication of their *Recommendations for a General Mathematical Sciences Program.* [3]

This set of recommendations, perhaps more than any other document, began a discussion and a gradual change in both the curriculum and the daily conduct of affairs in collegiate departments dealing with the discovery, development, and teaching of mathematics across the country. In fact, it was the major force in many departments changing their name from the Department of Mathematics to the Department of Mathematical Sciences. This was evidence that individuals, both within and outside the discipline, were coming to see mathematics as encompassing the traditional topics of calculus, algebra, geometry, analysis,...; the subject matter of probability, optimization, data analysis, and statistics; the applications of computer science; and the methods of modeling and of operations research. This reconceptualization of mathematics called for new coursework in discrete methods, applied algebra, and numerical analysis.

Recommendations for collegiate major programs were built on a philosophy that included foci on student attitudes about the mathematical sciences and the development of reasoning skills, approaches that featured interactive teaching styles incorporating student discovery and the study of applications present in project format. As such, the projected reforms in undergraduate mathematics were strong precursors to the National Council of Teachers of Mathematics' (NCTM) *Curriculum and Evaluation Standards for School Mathematics.* [8]

Central to the content recommendations was the call for the reworking of the core set of program requirements for an individual specializing in the mathematical sciences. This core included three semesters of calculus, linear algebra, probability and statistics, discrete methods, differential equations, some computing, and additional coursework drawn from abstract algebra, advanced calculus, and modeling/operations research. Most importantly, it helped modify a single listing of required coursework for the first two years that focused only on calculus and linear algebra and opened the way for integrating the contents of the seven core courses into a four-course unified program providing a broad range of options for students. The integration and highlighting of contents from calculus, linear algebra, differential equations, discrete mathematics, and probability and statistics has, in turn, increased the emphasis on applications and project-oriented problem solving.

Continued, and more focused, support for such programs was provided by later CUPM reports ([4], [16]), and other projects ([1], [5], [11], [12], [14], [15], [18]), as they released additional calls for a reform of the collegiate curriculum in the mathematical sciences centered in a reworking of the calculus, a creation of supporting courses reflecting the intra-disciplinary nature of mathematics itself, and a highlighting of interdisciplinary applica-

[1]Illinois State University, Normal, IL 61790

tions of the concepts and procedures of mathematics to quantitative and qualitative considerations in other fields of human activity. During the same period, NCTM produced its Standards for both the content of the curriculum and the ways in which students and teachers should come to understand mathematics ([8], [9], [10]). Additional documents, such as the MAA's *Heeding the Call for Change: Suggestions for Curricular Action* ([17]), provided suggestions for the directions that undergraduate programs might undertake to promote change. Sessions at annual meetings of the MAA, AMS, AMATYC, and NCTM dealing with the undergraduate curriculum, its teaching, and its problems filled to overflowing, indicating the interest and support of active professionals in the nation's colleges and universities.

However, few working models of the proposed programs have been constructed that really integrate the various areas of mathematics composing the recommended core; that feature the threads of reasoning, communication, problem-solving, technology, data analysis, and the historical roots of mathematics; or that weave these threads into a coherent set of offerings aimed at a broad range of students specializing in the mathematical sciences. It is not altogether clear why this is the case. Is it a matter of institutional rigidness, of faculty incapable of making such changes, of a lack of resources and support, of programmatic issues such as transfer students, or a fear of setting aside traditional paths for untested new paths?

The press of these recommendations is perhaps strongest in large institutions offering a broad range of majors (engineering, the sciences, economics and finance, teacher education,...) each requiring specific coursework for unique needs or discipline certification/licensure demands. As each of these partner discipline areas continue to "ratchet up" their use of quantitative concepts and procedures, they increase the need for more, and more varied, content to be included in the core offerings in the mathematical sciences. Historically, this demand for inclusion has been greatest in the first two years of the collegiate experience.

How can colleges and universities, from the two-year level to the major research institutions, meet this increasing demand? Few working models have been constructed that really integrate the topics often mentioned for such a core. At the present, many programs are struggling to meet yet another demand - shaping courses in broad quantitative reasoning for the liberal arts/general education core requirements for non-majors. Many of these same threads: reasoning, communication, problem-solving, technology, data analysis, and the historical roots of mathematics also need to be included in a comprehensive treatment of the core courses for those majoring in mathematics or one of its partner disciplines.

Given these needs, it is imperative that the mathematical sciences consider new ways of meeting the needs of both their majors entering new fields of endeavor and better serving their students coming from other allied disciplines. To do so, the mathematical sciences have to think of change in their curricula that extend beyond the calculus level. What are the core learnings that one would want all students to take with them from a consideration of calculus, of linear algebra, of discrete mathematics, of differential equations, of probability and statistics? How can these core learnings be fashioned into new programs of study for undergraduate students, especially undergraduate students enrolled in the first two years of their collegiate experiences so that they might make fruitful applications of these core learnings in their work in partner disciplines in the last two years of the undergraduate experience?

To answer such questions, several barriers must be considered. First is the fear among mathematics departments of losing control of their major. Is the purpose, even the future, of the mathematics department to serve the needs of other departments and, in the course of doing so, to become the "handmaiden" of other disciplines abandoning the service of those who would choose mathematics as their majors? Is it the fear among some mathematicians that doing so brings them into teaching in areas for which they were not prepared - applied mathematics? Is it the transfer/articulation problem that brings students into our programs mid-stream with different initial experiences? Is it institutional rigidness, of faculty incapable or unwilling to make such changes, or a lack of institutional resources and support for equipment and faculty development, or a fear of further migration of students from mathematics that serve to forestall change. All of these questions are important and serve as examples of the barriers to reshaping the undergraduate major to energize the mathematical sciences major while adequately serving the partner disciplines' prerequisite needs in the first two years.

One program stands out as a test bed for such changes. However, several features of its student body, its goals, and the ability of the institution to make sweeping changes almost overnight cause it to be viewed as a singularity. The United States Military Academy (USMA) at West Point has, since 1991, required all of its students to complete a four course mathematical sciences core integrating the content of three semesters of calculus and one semester each of discrete mathematics, linear algebra, differential equations, and probability and statistics. This seven courses into four model, or 7–into–4, has been successfully taught to thousands of students with positive reactions from both faculty in the mathematical sciences and the partner disciplines from engineering, the physical sciences, and economics ([19]). However, it is far easier for USMA to make such changes. It is not hampered by external influences such as transfer students, branch campuses, disconnected colleges and universities, student work scheduled, and other forces that make change, both in curriculum content and pedagogical approaches, difficult. Further, the Academy is a tightly bound and highly interconnected campus with a tremendous *esprit-de-corps* that has been molded over almost two centuries.

Given the needs for change cited and the existence of a test case at USMA, an NSF sponsored conference was held at the Academy on April 23-24, 1994 to consider the questions of core requirements for the courses often employed by partner disciplines, to consider what concepts and procedural skills are really central, and to consider how these questions might lead to the creation of a new undergraduate set of core requirements that meet both the needs of mathematics departments and those of their partner disciplines. These charges were shaped for the 60–plus participants as follows:

- What content and student growth goals would you propose for a core mathematics program in the first two years of collegiate study?
- What calculus, linear algebra, differential equations, probability and statistics, and discrete mathematics topics would you include in your program? How would you structure the content in these subjects, recognizing that there are other courses further downstream and in other departments relying on students' learning in these courses?
- How can aspects of a student growth model be integrated into the model you would propose?
- What implications does your model have for secondary schools, for upper division courses in mathematics, and for allied disciplines having mathematics requirements?

The chapters in the Sections I and II capture the focus papers presented, the comments of the reactors, and the general tenor of the conference. The two days of discussion were intense, fruitful, and perhaps far-reaching in their vision for what the next steps in the reform movement might be for U.S. schools, colleges, and universities as they strive to meet both process and content recommendations for change in mathematics teaching and learning the core curriculum areas of undergraduate mathematics.

In April, 1995, the Academy again played host to teams of faculty from a small number of invited institutions who were working on changing their core programs at the undergraduate level. Section III contains working plans for four of these institutions. These plans indicate beginning steps to revise the undergraduate major along the lines of a core curriculum serving many needs in the first few years, while still maintaining a solid basis for those students continuing to major in the mathematical sciences. These plans may foreshadow the shape of curricula of the future.

References

[1] Albers, D., Rodi, S., and Watkins, A. (eds.). (1985). *New Directions in Two-Year College Mathematics.* New York, NY: Springer-Verlag.

[2] Committee on the Mathematical Sciences in the Year 2000. (1991). *Moving Beyond Myths: Revitalizing Undergraduate Mathematics.* Washington, DC: National Academy Press.

[3] Committee on the Undergraduate Program in Mathematics. (1981). *Recommendations for a General Mathematical Sciences Program.* Washington, DC: Mathematical Association of America.

[4] Committee on the Undergraduate Program in Mathematics. (1991). *The Undergraduate Major in the Mathematical Sciences.* Washington, DC: Mathematical Association of America.

[5] Douglas, R.G. (ed.). (1986). *Toward a Lean and Lively Calculus.* MAA Notes #6. Washington, DC: Mathematical Association of America.

[6] Glimm, J.G. (ed.). (1991). *Mathematical Sciences, Technology, and Economic Competitiveness.* Washington, DC: National Academy Press.

[7] Madison, B.L., and Hart, T.A. (1990). *A Challenge of Numbers: People in the Mathematical Sciences.* Washington, DC: National Academy Press.

[8] National Council of Teachers of Mathematics. (1989). *Curriculum and Evaluation Standards for School Mathematics.* Reston, VA: Author.

[9] National Council of Teachers of Mathematics. (1991). *Professional Standards for the Teaching of Mathematics.* Reston, VA: Author.

[10] National Council of Teachers of Mathematics. (1993). *Assessment Standards for School Mathematics.* Reston, VA: Author.

[11] Ralston, A., and Young, G.S. (eds.). (1983). *The Future of College Mathematics.* New York, NY: Springer-Verlag.

[12] Roberts, A.W. (ed.). (1996). *Calculus: The Dynamics of Change.* MAA Notes #39. Washington, DC: Mathematical Association of America.

[13] Sigma Xi. (1990). *Entry-level Undergraduate Courses in Science, Mathematics and Engineering: An Investment in Human Resources.* Research Triangle Park, NC: Author.

[14] Solow, A. (ed.). (1994). *Preparing for a New Calculus.* MAA Notes #36. Washington, D.C.: Mathematical Association of America.

[15] Steen, L.A. (ed.). (1989). *Calculus for a New Century: A Pump, Not a Filter.* MAA Notes #8. Washington, DC: Mathematical Association of America.

[16] Steen, L.A. (ed.). (1989). *Reshaping College Mathematics.* MAA Notes #13. Washington, DC: Mathematical Association of America.

[17] Steen, L.A. (ed.). (1992). *Heeding the Call for Change: Suggestions for Curricular Activity.* MAA Notes #22. Washington, DC: Mathematical Association of America.

[18] Tucker, T.W. (ed.). (1990). *Priming the Calculus Pump: Innovations and Resources.* MAA Notes #17. Washington, DC: Mathematical Association of America.

[19] West, R. D. *Evaluating the Effects of Changing an Undergraduate Mathematics Core Curriculum which Supports Mathematics–Based Programs.* Unpublished doctoral dissertation, New York University.

Contents

Core Curriculum in Context: History, Goals, Models, Challenges
Lynn Arthur Steen 3
Response to Core Curriculum in Context: History, Goals, Models, Challenges
Joan Ferrini-Mundy 15
Core Mathematics at the United States Military Academy: Leading into the 21st Century
David C. Arney, William P. Fox, Kelley B. Mohrmann, Joseph D. Myers, and Richard A. West 17
Discrete Mathematics in the Core Curriculum
Alan Tucker 33
Response to Discrete Mathematics in the Core Curriculum
Martha Siegel 36
Calculus in the Core
Wayne Roberts 37
Response to Calculus in the Core I: Parallels in Calculus
David Heckman 43
Response to Calculus in the Core II: Calculus and Other 7-Into-4 Issues
Stephen Rodi 45
Calculus after High School Calculus
Don Small 47
Response to Calculus After High School I
Frank Wattenberg 57
Response to Calculus after High School Calculus II
Jeanette Palmiter 59
Linear Algebra in the Core Curriculum
Donald R. LaTorre 63
Response to Linear Algebra in the Core I
David Carlson 67
Response to Linear Algebra in the Core II
Anita E. Solow 69
Response to Linear Algebra in the Core III
Steve Friedberg 71
Differential Equations in the Core Curriculum
James V. Herod 75
Response to Differential Equations in the Core I
Donald Bushaw 85
Response to Differential Equations in the Core II: ODEs Renewed
Courtney Coleman 87
Response to Differential Equations in the Core III: Data Sets in the First Course of Differential Equations
David O. Lomen 89

Probability and Statistics in the Core Curriculum
David S. Moore 93
Response to Probability and Statistics in the Core I
Ann E. Watkins 99
Response to Probability and Statistics in the Core II
R. A. Kolb 101
Probability and Statistics in the Core III: Is It Really Such A Tight Squeeze?
Sheldon P. Gordon 103
Evaluation of Mathematics Core Curriculum Conference, April 22–24, 1994
Robert Orrill 109
A Curriculum Reform Workshop/Retreat
Don Small 115
Carroll College Mathematics Curriculum Reform Project
John Scharf and Marie Vanisko 117
A Preliminary Plan for Curriculum Change at Harvey Mudd College: n-into-four
Robert Borrelli, Robert Keller, Michael Moody 121
Core Curriculum Reform Model for the Oklahoma State University
James Choike 125
Plan to Integrate Calculus II and III Curricula at University of Redlands
J. Beery, P. Cornell, A. Killpatrick, and A. Koonce 130

Section I
Introduction to the Core Curriculum

Core Curriculum in Context
History, Goals, Models, Challenges

Lynn Arthur Steen[2]

As the designated historian for this consideration of the undergraduate mathematics major, I chose to begin forty years ago at the founding of the MAA Committee on the Undergraduate Program in Mathematics (CUPM). One could begin earlier, for as you will see from the record, the issues we have come here to discuss are not new—neither now, nor in 1953 when CUPM was founded. Questions about integrating the mathematics curriculum—or any curriculum—are as perennial as any debate in the history of education.

CUPM was established in 1953 by the Mathematical Association of America "to modernize and upgrade" the mathematics curriculum and to halt "the pessimistic retreat to remedial mathematics." (Actually, at the time of its founding, the committee was known merely as the Committee on the Undergraduate Program, or CUP.) In 1953, the total college enrollment in mathematics courses was 800,000; bachelor's degrees in mathematics numbered 4,000; Ph.D.'s, 200. During the intervening four decades, there has been a four-fold increase in these numbers—to three million mathematics students, 16,000 majors, 1000 Ph.D.'s. Most of that growth occurred during the first half of this period in the ramp-up of sciences and engineering during the post-Sputnik panic.

You'll be pleased to learn that CUPM's first project was a proposal for a course whose motivation and character were quite similar to the themes on our agenda for this weekend. Called *Universal Mathematics*, this course was intended as an integrated introduction to higher mathematics that provided in the first semester continuous mathematics, and in the second, discrete mathematics:

- First Semester: Analysis, College Algebra, Introduction to Calculus
- Second Semester: Mathematics of Sets and Elementary Discrete Mathematics

However, despite CUPM's urging, nothing much changed. Like similar proposals in more recent years, CUPM's new course was too radical. It emerged at a time when the typical curriculum at even the best institutions reflected an entrenched pattern that was by then several decades old:

- First Year: Trigonometry, College Algebra, Analytic Geometry
- Second Year: Differential and Integral Calculus
- Third Year: Advanced Calculus
- Fourth Year: Differential Equations, Theory of Equations,...

This early foray into curricular reform contains an important lesson for others who may have similar ideas: curriculum proposals from committees succeed primarily when they are a synthesis of evolving consensus, not the leading edge of radical reform.

To understand the context in which the early CUPM functioned, we must recognize important parallel events that helped shape the climate for the first two years of college mathematics in the 1950s and 1960s. Foremost was SMSG, a campaign led by mathematicians and mathematics educators to improve school mathematics. Then there were seminal university texts: Thomas in calculus, Birkhoff & MacLane in modern algebra. Other factors, particularly the emergence of computing and the expanding horizons of applications, forced the mathematics community to confront the reality of a new context for the undergraduate mathematics program. Thus began the tradition of curriculum reports that have been the landmark not only of CUPM, but also of the discipline of mathematics. No other subject has as strong a tradition of systematic nationwide examination of undergraduate curriculum as has mathematics. Table 1 gives a sample of titles to illustrate the steady flow of reports that began in the early years of CUPM.

A General Curriculum for Mathematics

In 1965 the National Academy of Sciences issued a report entitled *The Mathematical Sciences* that introduced a new umbrella designation to encompass the related disciplines of mathematics, statistics, operations research, and computer science. It also introduced the phrase "core mathematics" as an alternative to the controversial moniker "pure math-

[2]St. Olaf College, Northfield, Minnesota 55057

Four Decades of Curriculum Reports
(Selected MAA Publications)

1965	*A General Curriculum in Mathematics for Colleges*
1965	*Pre-Graduate Preparation of Research Mathematicians*
1971	*An Undergraduate Program in Computational Mathematics*
1972	*Introductory Statistics without Calculus*
1972	*Undergraduate Mathematics Courses Involving Computing*
1981	*Recommendations for a General Mathematical Sciences Program*
1986	*Toward a Lean and Lively Calculus*
1987	*Calculus for a New Century*
1989	*Discrete Mathematics in the First Two Years*
1991	*A Call for Change: The Mathematical Preparation of Teachers of Mathematics*
1992	*Perspectives on Contemporary Statistics*

Table 1

ematics" to more accurately describe those parts of the mathematical sciences that have traditionally occupied the intellectual center of the discipline and of the curriculum.

That same year, CUPM issued its first comprehensive curriculum report: *A General Curriculum in Mathematics for Colleges.* In this slim yet influential volume, CUPM described a single curriculum that can serve equally well as a foundation for students with varieties of interests and for colleges with varieties of missions. The report began with a sober analysis of the challenges posed by the then-current curriculum:

> Many of us can well remember when a firm tradition decreed that college mathematics should consist of the sequence: College Algebra, Trigonometry, Analytic Geometry, Differential Calculus, Integral Calculus, Differential Equations, Theory of Equations, Advanced Calculus,... That old structure, which comfortably regulated college mathematics, has fallen apart. (CUPM, 1965, p. 4)

This CUPM report goes on to diagnose, in terms that sound uncannily contemporary, the causes of this collapse and the challenges it poses:

> The generally welcome revolution in school mathematics... has created a greater diversity of entering students than we ever experienced before... and it has only begun. The *spread* in the mathematical capability of entering students will become much greater still.
>
> There are now many more kinds of mathematical knowledge... brought about by the computer, the increasing mathematization of the biological, management, and social sciences, and by the modern emphasis on such subjects as probability, combinatorics, logic,...
>
> Thus there is a multiple output as well as a multiple input to the mathematics department's 'black box'. (CUPM, 1965, p. 4)

It's specific curriculum recommendations are summarized in Table 2.

Among its many specific recommendations, the 1965 CUPM report

- Begins with elementary calculus based on prior study of elementary functions.
- Introduces multivariable calculus in the first year.
- Recommends linear algebra at the beginning of the sophomore year.
- Advocates introducing all students to "the type of algorithmic approach that enables a problem to be handled by a machine."
- Provides for multiple entry points, and multiple exits. Offers "a more suitable compromise between the whole of calculus and no calculus than does the conventional course structure."
- Economizes offerings to enable the entire major to be taught by a four person department.
- Starts intuitively, with increasing rigor. "Start where the student really is, and proceed to where he [or she] should be."
- Includes options for one-year mathematics requirements: Math 0-1, or 1-2P.

Reflections

By 1972, CUPM reflected on what it had tried to accomplish in this first attempt at standardization of the undergraduate mathematics major. It did

A General Curriculum in Mathematics for Colleges

Lower Division Courses:

0. *Elementary Functions.* Polynomials, rational, algebraic, exponential, and trigonometric functions.
1. *Introductory Calculus.* Differential and integral calculus of elementary functions with associated analytic geometry.
2. *Mathematical Analysis I.* Techniques of one and several variable calculus, series.

2P. *Probability.* Sample spaces, random variables, limit theorems, statistical inference.

3. *Linear Algebra.* Linear equations, vectors spaces, linear mappings, matrices, quadratic forms, eigenvalues, geometric applications.
4. *Mathematical Analysis II.* Multivariable calculus, linear differential equations.

Upper Division Courses:

5.	Advanced Multivariable Calculus.	8.	Numerical Analysis.	11.	Real Variable I.
6.	Algebraic Structures.	9.	Classical Geometry.	12.	Real Variable II.
7.	Statistical Theory and Inference.	9'.	Differential Geometry.	13.	Complex Analysis.
7'.	Probability and Stochastic Processes.	10.	Applied Mathematics.		

Table 2

this by publishing a "Commentary" on the 1965 report in which it raised explicitly the question of formally shifting the focus of departments from "mathematics" to "the mathematical sciences": "When thinking about undergraduate education, is it not now more appropriate to speak of *the* mathematical sciences in a broad sense rather than simply of mathematics in the traditional sense?" (p. 2).

The 1972 *Commentary* reflected on the criticisms that had come to the Committee in the intervening years:

- The pace of the course outlines was too fast.
- The syllabi leave no room for applications.
- Departments also have other substantial commitments.
- Mathematics is broadening beyond its classical boundaries.
- Some traditional core courses are less relevant than newer applied courses.
- The lower-division service responsibilities of a mathematics department are not well met by a single calculus-based track.
- The increasingly diverse mathematical preparation of entering students requires multiple points of entry prior to calculus.

The *Commentary* concludes with an observation that virtually recants the title of the 1965 CUPM report: "Thus... it is no longer clear that there should be a single general curriculum in mathematics." (p. 3)

The *Commentary* contains a number of explicit recommendations that extend or revise what CUPM had proposed in the 1965 *General Curriculum* report. These include:

- An explicit retreat to the tradition of a basic two-semester calculus course during the first year of college mathematics. "CUPM does not wish to suggest any alternative for the first year of calculus."
- A recommendation that each exit should be logical and coherent. Math 1 should be a self-contained introduction to one-variable calculus, including the Fundamental Theorem of Calculus; Math 1-2 should be a self-contained introduction to the ideas of calculus of both one and several variables, "including the first ideas of differential equations."
- A reaffirmation of the placement of Linear Algebra (Math 3) in third semester. Argues for making Math 4 build more explicitly on the concepts from Math 3.
- A suggestion that Math 5—the fourth term of traditional calculus (vector calculus and Fourier methods)—is no longer necessary in the core for all mathematics majors.
- An urging that Introductory Modern Algebra (Math 6M) be part of the core for all mathematics students, and that colleges offer a sequel (Math 6L) that provides more advanced linear algebra.

Mathematics vs. The Mathematical Sciences

It is clear from both the 1965 and 1972 reports that CUPM's general philosophy for lower division mathematics was unambiguously designed to broaden the scope of mathematical exposure of students during their first two years in college: not only calculus, but also probability and linear algebra were recommended for these first two years. The

Pre-Graduate Preparation of Research Mathematicians
(A Sample of Topics from a 1965 CUPM Report)

First year:	Second year:
Tangents, derivatives, parametric curves	Characteristic equation and eigenvalues
Integral as positive linear functional	Cayley-Hamilton theorem; Gram-Schmidt process
Proof that cont. functions are integrable	Open, closed, compact sets
Pointwise and uniform convergence	Completeness of space of continuous functions
Monotone convergence theorem	Integral as uniformly continuous positive linear functional
Linear transformations	Implicit function theorem
Vectors in two and three dimensions	Differential p-forms
Rank and nullity theorem	Divergence, curl, Stokes' and Gauss' theorems

Table 3

move towards "the mathematical sciences" provided a philosophical context in which these recommendations were embedded.

However, this consensus of movement from mathematics as traditionally defined to the mathematical sciences broadly conceived was not the only force influencing undergraduate mathematics. Other CUPM panels in this same era produced a series of recommendations for students intending to pursue advanced study of mathematics that were rooted more in the spirit of Bourbaki, in which the central core of mathematics was a beautiful intellectual synthesis of analysis, geometry, and algebra.

The CUPM Panel on Pre-Graduate Preparation of Research Mathematicians, writing in 1965, offered an ideal program (see Table 3) which was, as they admitted, "somewhat unrealistic," but nonetheless "suitable for honors programs" and "as a goal for the regular curriculum." Theirs was an explicitly mathematical vision emphasizing from the very beginning the intrinsic unity of the subject. "Calculus should be presented so as to introduce and utilize significant notions of linear algebra and geometry in the construction of analytic tools for the study of transformations of one Euclidean space into another... Material should be arranged and presented in such a manner that students are ever mindful of mathematics as an interrelated whole rather than as a collection of isolated disciplines." (pp. 4–5)

Computational Mathematics

In 1971, anticipating the monumental changes that computers were going to bring to the practice of mathematics, CUPM issued a special set of recommendations for an undergraduate program in computational mathematics that was "intended to be a departure from the traditional undergraduate mathematics curriculum." This report, rooted in very different assumptions than those of the pregraduate preparation report, recognized the need for "innovative undergraduate programs that provide for a wide range of options, for different opportunities for graduate study, and for a variety of future careers." (p. 8) The CUPM program in computational mathematics was intended to be one of "several equally valid options" for students in the mathematical sciences. By building on the 1965 *General Curriculum* this new program was specifically designed to "permit continuation in computational mathematics or in pure mathematics, with suitably selected senior courses." (p. 8)

Table 4 shows the list of courses recommended for this program: twelve courses in mathematics (M1-M5), computing (Cl-C3), and computational mathematics (CM1-CM4) to be taken during the first three years of undergraduate study. The mathematics courses are computationally slanted versions of the courses previously recommended by CUPM; the computer courses are versions of those recommended by the Association of Computing Machinery (ACM) in "Curriculum '68" (later revised in "Curriculum '78"). The four courses in computational mathematics represent a new hybrid intended to build substantive links between computing and mathematics.

Statistics and Discrete Mathematics

Continuing its campaign to move mathematics toward the mathematical sciences, in 1972 CUPM published recommendations for *Introductory Statistics without Calculus.* The purpose of such a course was primarily to introduce the ideas of variability and uncertainty, and—contrary to tradition—only secondarily to introduce standard formulas,

An Undergraduate Program in Computational Mathematics 2
(1971 Recommendations of CUPM)

Ml.	Introductory Calculus.	CM1.	Computational Models & Problem Solving.
M2.	Mathematical Analysis I.	CM2.	Introduction to Numerical Computation.
M3.	Linear Algebra.	CM3.	Combinatorial Computing.
M4.	Mathematical Analysis II.	CM4.	Diff. Equations and Numerical Methods.
M5.	Advanced Multivariable Calculus.		
Cl.	Introduction to Computing.	Year 1:	Ml, M2, Cl, CMl
C2.	Computer Organization and Programming.	Year 2:	M3, M4, C2, CM2
C3.	Programming Languages and Data Structures.	Year 3:	M5, C3, CM3, CM4

Table 4

terms, and techniques. This first course in statistics, CUPM argued, should emphasize inferential concepts and data analysis, not mathematical elements.

CUPM recommended several types of courses to achieve this objective. Each would employ real data sets and build on the computational power of computers for simulation, calculation, and interactive learning. These recommendations emphasize hands-on methods, including demonstration experiments (e.g., tossing a coin 100 times) that illustrate predictable patterns; open experiments (e.g., tossing thumbtacks to see how they land) that illustrate empirical phenomena whose patterns cannot readily be predicted *a priori*; and simulations (e.g., of queues and servers) of patterns that can best be observed through computer methods.

The main-stream recommendation in this report describes a fairly traditional course, but CUPM added three rather different alternatives to stress the point that it was statistical thinking, not just a collection of formulas, that was the goal of this recommendation:

S1. *Elementary Statistics.* Statistical description; probability, random variables; probability distributions; sampling distributions; inferences about population means.

S2. *Decision Theory.* Bayes' strategies; significance levels; confidence intervals.

S3. *Nonparametric Statistics.* Chi-square; contingency tables; correlation; robustness.

S4. *Case Studies.* Modeling; computer simulation; open-ended, meaningful problems.

The message about the nature of introductory statistics seems to have a hard time being heard, or implemented. Twenty years after this 1972 report, a recent MAA volume entitled *Perspectives on Contemporary Statistics* asserts that "a wide gap separates statistics teaching from statistical practice." It recommends yet again, echoing the 1972 report, that instruction in statistics should emphasize data: analyzing data, producing data, and inference from data.

Also in 1972, in response to the growing importance of computing, CUPM published recommendations intended to nudge the content of mathematics courses in the direction of computational procedures by stressing algorithms, approximations, model building, and problem-solving processes. Four course outlines (MC0-MC3) offer computer-oriented versions of M0-M3. One new course, Discrete Mathematics (DM), is introduced both to supplement the standard curriculum and to "serve well as a first mathematics course for students from many disciplines."

DM *Discrete Mathematics.* Set theory, permutations and combinations, pigeonhole principle, generating functions, difference equations, relations, graphs, circuits, paths, Eulerian and Hamiltonian paths, network flow problems.

So by 1973, the mid-point of CUPM's history, we find not a "7 into 4" problem but what amounts to a "20 into 4" problem. Between 1965 and 1973, CUPM had proposed twenty different introductory courses in mathematics, computing, and statistics, each with legitimate claims for being part of the mathematical repertoire of any serious student intending to study a mathematics-intensive field (see Table 5).

A General Mathematical Sciences Program

In 1981 CUPM issued a new report entitled *Recommendations for a General Mathematical Sciences Program* that set firmly in place the notion that

CUPM Recommendations as of 1973

M1.	Trigonometry and Algebra.	CM1.	Computational Models and Problem Solving.
M0.	Elementary Functions.	CM2.	Introduction to Numerical Computation.
M1.	Introductory Calculus.	CM3.	Combinatorial Computing.
M2.	Mathematical Analysis I.	CM3.	Combinatorial Computing.
M2P.	Probability.	CM4.	Differential Equations and Numerical Methods.
M-DM.	Discrete Mathematics.		
M3.	Linear Algebra.		
M4.	Mathematical Analysis II.		
M5.	Advanced Multivariable Calculus.		
C1.	Introduction to Computing.	S1.	Elementary Statistics.
C2.	Computer Organization and Programming.	S2.	Decision Theory
C3.	Programming Languages and Data Structures.	S3.	Nonparametric Statistics.
		S4.	Case Studies.

Table 5

mathematical sciences rather than mathematics is the proper subject of the undergraduate program. "CUPM now believes that the undergraduate major offered by a mathematics department at most American colleges and universities should be called a Mathematical Sciences major."

In this report, CUPM argued that first courses should appeal to "as broad an audience as is academically reasonable." "The mathematical science curriculum should be designed around the abilities and needs of the average student, with supplementary work to attract and challenge talented students." To broaden the appeal of mathematics—at a time when the number of majors had dropped by over 50% from its historic high—CUPM argued that computer science, applied probability, and statistics should be "an integral part of the first two years of college mathematics."

As part of this report, CUPM recommended three new courses that should fit into the first two or three years of the undergraduate curriculum:

Discrete Structures. Combinatorial reasoning (graph theory, combinatorics) taught at the level of introductory calculus.

Applied Algebra. Sets, partial orders, Boolean algebra, finite state machines, formal languages, semigroups, modular arithmetic, automata, enumeration theory, lattices. (Adapted from ACM's "Curriculum '68" and "Curriculum '78.")

Statistical Methods. A post-calculus course emphasizing data (organization and description), probability (random variables, distributions, Law of Large Numbers, Central Limit Theorem), and statistical inference (significance tests, point estimation, confidence intervals, linear regression).

The most contentious issue in this 1981 report concerned not the first two years of the undergraduate program, but the last two. Recognizing the message being sent by students who had "voted with their feet," CUPM dropped the historical requirement of year-long courses in both algebra and real analysis as the upper division core of the mathematics major. They recommended, instead, at least one rigorous two-course sequence at the advanced level—which might be, for instance, in applied mathematics or in probability and statistics. This recommendation recapitulated at the advanced level the contrast revealed fifteen years earlier by the two very different CUPM proposals for the content of the first two years of college mathematics.

Discrete Mathematics

The 1980's witnessed a flurry of curricular exploration related to discrete mathematics, whose importance grew in proportion to the increasing demand for computer science. Discrete mathematics was to be the language of the information age, as calculus had been the language of the age of (Newtonian) science. The key curricular issue for mathematicians was whether it was possible to design a year-long course in discrete mathematics that could hold its own in head-to-head competition with calculus as a legitimate entry point for the study of college mathematics. While most institutions experimented with separate courses, a few tried to devise an integrated approach approximately in the spirit of CUPM's stillborn *Universal Mathematics* of the mid-50's.

The Enrollment Facts of Life
(1990–91 CBMS Survey)

	2-yr	4-yr	Total
Remedial	724	260	984
Algebra & Trig	245	360	605
Finite/Business/Lib. Arts	90	173	263
Elem Sch Teach	9	62	71
Technical Mathematics	18		18
Precalculus Level			**1941**
Calculus	124	545	669
Differential Equations	4	40	44
Discrete Mathematics	1	17	18
Linear Algebra	3	42	45
Calculus Level			**776**
Elementary Statistics	47	87	134
Elementary Probability	7	32	39
Statistics & Probability			**173**
Comp & Society	10	69	79
Packages	21	73	94
CS1	32	80	112
CS2	12	23	35
Other Elem Comp. Sci	23	86	109
Computer Science			**429**
Total Elementary Courses			**3319**

Table 6

At least three volumes published during the 1980's record the recommendations and experiences of individuals who led these experiments. A special committee of the MAA concluded a study of this movement with the clear recommendation that "discrete mathematics should be part of the first two years of the standard mathematics curriculum at all colleges and universities, and should be taught at the intellectual level of calculus." They provided a rather standard course description:

> *Discrete Mathematics.* Algorithms, graph theory; combinatorics, induction, recurrence relations, difference equations, logic, introductory set theory

and outlined two options for implementation:

- Two one-year sequences in discrete mathematics and in (streamlined) calculus;
- A two-year integrated course in discrete and continuous mathematics (calculus).

Students and Courses

Curriculum recommendations that concentrate on calculus, statistics, discrete mathematics, linear algebra, and differential equations—the classic "7 into 4" problem—ignore the most important reality: students. It turns out, if one looks at the data (Figure 6), that most students take other courses—primarily those that are part of the traditional high school curriculum. Only one in three beginning college mathematics enrollments is in any of the seven courses we will be discussing. So in planning these core courses, we should not fall into the trap of assuming either that all able students are in one of these courses, or that all mathematics courses suitable for beginning students are included among the seven on our list The reality of matching students and courses is far more complex: mathematics students do not simply enroll in courses, but are created by courses.

Goals and Objectives

Because there is so much overlap in both content and context between the mathematics taught in high school and the mathematical sciences covered during the first two years of college, it is useful to examine the goals of the curriculum from both school and college perspectives. The best known statement of contemporary goals can be found in the NCTM *Standards for School Mathematics*(1989). This influential document identifies five broad goals

Strategies for Success in Undergraduate Mathematics
(from diverse NCTM, AMATYC, and MAA reports)

- Teach in ways that engage students and encourage learning.
- Provide for the mathematical needs of all students.
- Engage technology in substantive support of mathematical practice.
- Blur early distinctions between majors and non-majors.
- Build smoothly on the standards-based core school curriculum.
- Motivate theory with applications.
- Provide extensive opportunities for students to read, write, listen, and speak.
- Help students learn how to learn mathematics.
- Recognize that content and pedagogy are inseparable.
- Use assessment to reinforce the goals of instruction.

Table 7

for mathematics education that define what NCTM terms "mathematical power":

- To reason mathematically
- To value mathematics
- To communicate mathematically
- To develop confidence
- To solve problems

In contrast, the following goals from a draft of standards being developed by the American Mathematical Association of Two Year Colleges (AMATYC) focus on the empowerment of students, what one might term "student power":

- *Empowerment*: Increase participation in mathematics-based careers by students heretofore under-represented in those fields.
- *Confidence*: Provide rich, deep experiences that encourage independent exploration, build tenacity, and reinforce confidence in each student's ability to use mathematics effectively.
- *Connections*: Present mathematics and science as a developing human activity that is richly connected with other disciplines and areas of life.
- *Citizenship*: Illustrate the power of mathematical and scientific thinking as a foundation for independent life-long learning.

Turning to the core curriculum itself, many MAA reports identify at least four distinct missions that must be served by introductory college mathematics:

- To ensure numeracy for all college graduates.
- To provide students with mathematical skills for further study and work.
- To prepare prospective teachers to implement a standards-based curriculum.
- To attract able students to major in the mathematical sciences.

All the standards recommendations—from NCTM, from AMATYC, from MAA—agree on certain strategies that are essential to achieve any measure of success in college mathematics. Unfortunately, these strategies (see Table 7) are honored more in rhetoric than in reality. Yet the considerable investment of recent years in reform of mathematics education has taught us many lessons about what works and what doesn't. As we embark on discussions about compressing seven courses into four, or reinventing the goals of introductory college mathematics, we would do well to heed some of these lessons from practice and research, from reform and accomplishment (see Table 8).

Models

It is not for me to talk about specific models: that is what this conference is all about. I would, however, like to set the context for your thinking about models by stressing the diversity of higher education—of institutions, of programs, of student goals, and of student preparation. Since students typically enroll in mathematics as part of a defined program of study, the context of these programs is of crucial importance to the nature and goals of introductory college mathematics.

Diversity of institutions and programs is one of the distinctive strengths of the American system of higher education. Students study mathematics in over 3000 post-secondary institutions of widely differing missions and purposes:

- Research Universities
- Comprehensive Universities
- Liberal Arts Colleges
- Community Colleges
- Vocational Institutes
- Employee Training Programs

Lessons Learned about What Works in Undergraduate Mathematics

From Practice:

- Learning takes place when
  - The student is enmeshed in community;
  - The subject is embedded in context;
  - The instruction is infused with inquiry.

From Research:

- Learning is construction of knowledge, and construction of knowledge is construction of motivation.
- Mathematical knowledge is not merely remembered, but is privately constructed, becoming unique to the individual.

From Calculus Reform:

- Writing to learn is as important as learning to write.
- Technology is the most powerful transforming agent.
- Mathematical models are tools for understanding.
- Group work encourages cooperative approaches.
- Multiple representations aid student understanding of mathematics.

From Successful Programs:

- View teaching as a collective responsibility of the entire department.
- Make teaching a public activity, supported by regular discussion and seminars.
- Form natural communities on educational issues as a context for peer review.

Table 8

Their programs of study are equally diverse, with widely different mathematical expectations depending on the particular purpose:

- Liberal Arts
- Business
- Humanities
- Education
- Pre-Law and Pre-Med
- Vocational
- Science and Engineering
- Computer Science

Table 9 provides a profile of student preparation as they begin their study of college mathematics. The first three columns display percentages of students prepared at different mathematical levels at three different stages of the school-college transition: upon graduation from high school, upon entrance to college, and upon enrollment in collegiate mathematics courses. One can see first that students leave high school with a considerable range of preparation in mathematics; that those who enter college are better prepared mathematically than the typical high school graduate; but that the range of demonstrated competence of college students based on college course enrollments is much weaker than the recorded preparation based on high school courses completed.

Challenge I: An Integrated Course

This conference is intended to address two basic questions. The first is about the content of the core: "How to integrate the critical content of calculus I, II, III, linear algebra, differential equations, probability and statistics, and discrete mathematics into a four-course sequence."

As we have seen, this question has a long history. My recital began nearly forty years ago with the CUPM proposal for a course in Universal Mathematics, but the vision of an integrated introductory course has deep roots in nineteenth century European curricula. In this era we have several contemporary experiments motivated by the same vision: The College Board's "Pacesetter" program provides an integrated modeling-based pre-calculus course, while COMAP's "The Foundation" provides a similar modeling-based integrated approach to first year college mathematics.

The challenge of these courses is to become mainstream. In contrast to many other college subjects (e.g., physics, chemistry, economics), mathematics has no tradition of beginning the undergraduate curriculum with a "Principles of Mathematics" course. This idiosyncrasy may be related to another distinctive characteristic of mathematics—that it is one of only two university subjects (the other being English) that builds in essential ways on a full K–12 curriculum. Students' mathematics education is in full swing by the time they enter college—which makes it difficult to develop an effective introductory course based on the ideal of "one-size-fits all."

The other more obvious reason that integrated introductions to mathematics have never succeeded is the controlling influence of engineering and its mathematical prerequisites. The tradition of begin-

Mathematical Preparation of Entering College Students

A. Highest Mathematical Course Completed by High School Graduates
B. Highest Mathematical Course Completed by College Students
C. Achievement Level Inferred from College Enrollments
D. Cumulative Course Completion of High School Graduates
E. Cumulative Course Completion of College Students
F. Cumulative Achievement Inferred from College Enrollments

A	B	C	Level	D	E	F
14%	0%	12%	Arithmetic	100%	100%	100%
15%	10%	22%	Algebra I	86%	100%	88%
20%	25%	18%	Geometry	71%	90%	66%
25%	30%	20%	Algebra II	51%	65%	48%
18%	25%	21%	Precalculus	26%	35%	28%
8%	10%	7%	Calculus 1	8%	10%	7%

Table 9

ning with a particular course—calculus—is largely due to the needs of engineering students in the major universities. But now that students take college mathematics for many other reasons as well, it is appropriate to take up the challenge of alternative courses. It is, however, more likely that evolution will dictate a bush-like structure to the curriculum (e.g., choices of entry points including calculus, discrete mathematics, statistics, and computing) rather than a new tree whose trunk is a planned, integrated course that serves as a substitute for calculus.

Challenge II: A Compelling Course

The second question posed by this conference is, in my judgment, more apt and more critical: "What is the role played by a set of fundamental core courses in launching the study of mathematics for students majoring in the mathematical sciences or in mathematically-dependent fields?" This question is interesting because it focuses on the student, not the subject. It turns our attention away from the aesthetics of curriculum design to the practical reality of how the curriculum can attract (or repel) prospective mathematics students. A moment's thought produces a host of difficult yet important questions:

- How do introductory courses influence student decisions about majoring in mathematics-intensive fields? (To succeed, students need to be welcomed into a community in which they can grow in confidence while supported by a safety net of friends and faculty who help them overcome mistakes and insecurity. In what ways can the curriculum, and its implementation, accomplish this important task of mathematical acculturation?)
- How can departments provide multiple entry points to diverse curricular paths that lead to productive careers? (Clearly, courses that lead to curricular dead-ends should not be offered. But how, for instance, can a student who begins with statistics and gets "hooked" move on to complete a major in the mathematical sciences in a natural progression that builds on the foundation of statistics rather than those of calculus?)
- How can course sequences be planned to serve well those, often the majority, for whom the course they are currently in will be their final mathematics course? (Ideally, each course must address the broad goals of mathematics education, and must leave students with a positive attitude about mathematics. Can this be done without compromising the "coverage" requirements of each course?)
- How can first year courses be organized to benefit equally students from standards-based school programs as well as those from traditional programs? (Students who enter college in coming years may come prepared for a robust, hands-on, modeling approach to mathematics; others will come with traditional expectations of a paper-and-pencil problem-oriented course; still others will come with chaotic mixtures of skills, accomplishments, and expectations.)
- How can departments ensure that different tracks all meet similar broad goals and ensure flexible future transitions for students who change career goals? (One strategy is to reduce prerequisites to a minimum and introduce specific background on an as-needed basis. That way students will be encouraged to work for broad objectives rather

than meeting narrow and oftentimes somewhat arbitrary prerequisites.)

Criteria

Finally, I leave you with three criteria by which to judge proposals for a core program. You may think these are mutually contradictory. But I claim that they are logically necessary if we are to succeed with undergraduate mathematics.

- Core courses should serve equally well all students in every course.
- Core courses should attract students to continue the study of mathematics.
- Core courses should launch new students into mathematics-intensive fields.

Response to Core Curriculum in Context: History, Goals, Models, Challenges

Joan Ferrini-Mundy[3]

It's very reassuring to know that others before us have tackled these challenging issues, and to assume that others will follow to continue efforts to understand these matters. But, I'd like to suggest some additions to our agenda for this meeting. I was glad to hear Lynn Steen's highlights about teaching that engages students and encourages learning. His remarks suggest a call for a simultaneous focus on issues of pedagogy and learning alongside the challenging matters of content order and course organization. I believe that mathematics content and pedagogy are inseparable. As someone who works in both the worlds of the high school and the university, it seems clear to me that the reforms will drive each other. Consider students arriving at institutions of higher education, coming from *Standards*-based (NCTM, 1989, 1991) secondary school programs. They may have experienced:

- alternative assessment practices (portfolios, performance assessments, writing to demonstrate mathematical knowledge, group assessments, open-ended tasks)
- classrooms where "discourse" is a central feature in mathematics learning (listening to, responding to, and questioning the teacher and one another in learning)
- active engagement with tools for learning mathematics (technology, manipulative materials)
- a shift away from the teacher as the sole authority for right answers, but rather the classroom as a mathematical community.

What kinds of mathematical understandings and processes are students gaining as they experience programs of this sort? How are we able to describe what they know and can do sufficiently to make a case for the changes we are proposing at the post-secondary level? How can the undergraduate experience build most effectively on their secondary school experiences?

There are additional *significant challenges* that we need to consider. Consider the following questions:

- How can curriculum changes benefit from and extend the knowledge base about student learning? Can proposed curriculum changes at the undergraduate level generate useful research about how students learn? (E.g.: what meaning does "multiple representations" have, for example, in areas of discrete mathematics? Can research guide us in determining issues of order of topic introduction, of emphasis, etc.?)
- How can focus on pedagogical issues be sustained during discussions of content reorganization?
- What experiences do faculty members need in order to become committed to these new directions?
- Do the mathematical sciences include mathematics education? I think that any well-educated mathematics professional should have knowledge of K–12 mathematics education issues, the fundamentals of mathematics learning, curricular trends, and about effective modes at communication in mathematics. Can this occur through a core curriculum?
- How do we think about prospective teachers within the proposed core curriculum? They will be asked to teach a *"Standards-based"* curriculum, or even more fundamentally, to lead the introduction of a "Standards-based" curriculum in schools. Will a core curriculum prepare them for this?

References

[1] National Council of Teachers of Mathematics (1989). *Curriculum and Evaluation Standards for School Mathematics.* Reston, VA: NCTM.

[3]University of New Hampshire, Durham, NH 03824

[2] National Council of Teachers of Mathematics. (1991). *Professional Standards for the Teaching of Mathematics.* Reston, VA: NCTM.

Core Mathematics at the United States Military Academy: Leading into the 21st Century

David C. Arney, William P. Fox, Kelley B. Mohrmann,
Joseph D. Myers, and Richard A. West [4]

The Department of Mathematical Sciences at the United States Military Academy is prepared to lead the young minds of America into the 21st century with a bold and innovative curriculum coupled with student and faculty growth models and interdisciplinary lively applications. In 1990, the mathematics department began its first iteration of their "7 into 4" core curriculum. Each year improvements have been incorporated into the core mathematics program. In 1992, interdisciplinary applications appeared in the core program as an opportunity to communicate and work with the academic disciplines. Our core curriculum is tied together both vertically and horizontally with threads. These threads tie together both the content within each course as well as among all the courses. Student attitudes are measured through course surveys as we attempt to develop "life-long learners". Student performance is measured or calibrated throughout their four years.
This article appeared in *PRIMUS*, December 1995, vol. V, no. 4, and is reprinted with permission of the editor of *PRIMUS*.

Introduction and Historical Perspective

The Department of Mathematical Sciences, USMA, has had a rich history of contributing to the education of our students as confident problem solvers and of developing its faculty as effective teachers, leaders, and researchers. The story of mathematical education at West Point is full of interest: faculty curriculum developments, teaching methods and tools, and technological equipment. Many of the department's advances have been exported to be utilized by other civilian and military educational institutions.

The actual teaching of mathematics at West Point dates back to before the Academy was established [7]. In 1801, George Baron taught a few Cadets of Artillery and Engineers some of the fundamentals and applications of algebra. The Academy at West Point was instituted by act of Congress and signed into law by President Thomas Jefferson on 16 March 1802. The first acting professors of mathematics were captains Jared Mansfield and William Barron. They taught the first few cadets algebra, geometry, and surveying.

Since the Academy was the first scientific and technical school in America, the early mathematics professors at the Academy had the opportunity to make significant contributions not only to the Academy, but also to other American colleges. Perhaps the most prominent contributors were the early 19th century department heads Charles Davies and Albert E. Church. The work of these two professors had a significant impact on elementary schools, high schools, and colleges across the country. Davies became a professor of mathematics in 1823. He was a prolific textbook author, writing over 30 books from elementary arithmetic to advanced college mathematics. His books were used in schools throughout the country. He had a tremendous influence on the educational system of America throughout the 19th century. Albert Church succeeded Davies in 1837, and served as Department Head for the next 41 years. Another influential author, he published seven college mathematics textbooks [7].

Our Department faculty lists notable military leaders for the country. Robert E. Lee was a standout student-instructor in the Department, Omar Bradley served as an Instructor for four years, Harris Jones and William Bessell were Deans of the Academic Board at USMA for a total of 15 years, and Department Heads Harris Jones, William Bessell, Charles Nicholas, John Dick, and Jack Pollin served

[4]U.S. Military Academy, West Point, NY 10996

impressively during two world wars [7].

The unique technical curriculum in place at the Academy during the middle of the 19th century produced many successful mathematicians and scientists for the country at large. West Point graduates Horace Webster, Edward Courtenay, Alexander Bache, James Clark, Francis Smith, Richard Smith, Henry Lockwood, Henry Eustis, Alexander Stewart, and William Peck filled positions as professors of mathematics or college presidents at other schools such as the U. S. Naval Academy, Geneva College, University of Virginia, University of Pennsylvania, University of Mississippi, Yale, Brown, Harvard, Columbia, Virginia Military Institute, Cooper Institute, and Brooklyn Polytechnic Institute. Two mathematics department heads became college presidents after leaving the Academy; Alden Partridge founded and became the first president of Norwich University, and David Douglas served as president of Kenyon College in Ohio for four years. Jared Mansfield was appointed surveyor-general of the Northwest Territory, and Ferdinand Hassler became superintendent of the United States Coastal Survey [7]. The West Point model of undergraduate mathematics education was spread throughout the nation by capable individuals such as these.

While the faculty at the Academy has been primarily military, the Department has benefited from civilian visiting professors since 1976. As part of the goal for civilianization of 25% of the faculty by 2002, begun in 1992, the Department established in 1994 a Center for Faculty Development in Mathematics. This Center will establish faculty development models and curricula and provide for the development of the "Davies Fellows", who serve as rotating civilian faculty members.

Sylvanus Thayer's first task before assuming the Superintendancy in 1817 was to tour the technical institutions of Europe and assess what features the Academy could use to advantage. One of Thayer's many accomplishments was to obtain numerous mathematics and science books from Europe. Thayer's book collection included many of the finest books available at that time. It provided a solid foundation for the Academy library to build upon. Today, the West Point Library has one of the finest collections of pre-20th century mathematics books in the world. Also during the middle of the 19th century, the Academy instructors used elaborate physical models made by Theodore Oliver to explain the structures and concepts of geometry [7]. This magnificent collection of string models is still in the Department today.

After Thayer studied the military and educational systems of Europe, he reorganized the Academy according to the French system of the Ecole Polytechnic. The Department of Mathematics faculty included as Professor the distinguished scientist and surveyor Andrew Ellicott, and the French mathematician Claude Crozet whom Thayer recruited during his European trip to bring to the Academy his expertise in Descriptive Geometry, advanced mathematics, and fortifications engineering. Combining the French theoretical mathematics program with the practical methods of the English, the Academy established a new model for America's program of undergraduate mathematics. This program of instruction in mathematics grew over several decades and was emulated by many other schools in the country. The initial purpose of the Military Academy was to educate and train military engineers. Sylvanus Thayer, the "Father of the Military Academy" and Superintendent from 1817-1833, instituted a four year curriculum with supporting pedagogy to fulfill this purpose. Thayer's curriculum was very heavy in mathematics; from Thayer's time to the late 1800's, cadets took the equivalent of 54 credit hours of mathematics courses. The topics covered in these courses were algebra, trigonometry, geometry, descriptive geometry (engineering drawing), analytic geometry, and calculus. Over the years, the entering cadets became better prepared and fewer of the elementary subjects were needed. During Davies' tenure (1823-37), calculus was introduced as a requirement for all cadets, and was used in the development of science and engineering courses. The time allotted for the mathematics curriculum decreased to 48 credit hours by 1940, and to 30 credit hours by 1950. During the 1940's, courses in probability and statistics and in differential equations were introduced into the core curriculum and a limited electives program was started for advanced students. In the 1960's, department head Charles Nicholas wrote a rigorous and comprehensive mathematics textbook that cadets used during their entire core mathematics program [7]. With this text, he was able to adapt the mathematics program to keep up with the increasing demands of modern science and engineering. In the 1970's, Academy-wide curricular changes provided opportunities for cadets to major in mathematics.

During the 1980's, a mathematical sciences consulting element was established that allowed faculty members and cadets to support the research needs of the Army. This type of research activity continues today in the Army Research Laboratory (ARL) Mathematical Sciences Center of Excellence and in the Operations Research Center (ORCEN). In 1990 the Department introduced a new core mathematics curriculum which included a course in discrete dynamical systems, with embedded matrix algebra. In that same year, the department changed its name to the Department of Mathematical Sciences to reflect broader interests in applied mathematics, operations research, and computation.

USMA has a long history of technological innovation in the classroom. It was Crozet and other professors at USMA in the 1820's who used the blackboard as the primary tool of instruction. In 1944, the slide rule was issued to all cadets and was used in all plebe mathematics classes. During William Bessel's tenure (1947-1959), the mathematics classrooms were modernized with overhead projectors and mechanical computers. Bessel was also instrumental in establishing a computer center at West Point. The hand held calculator was issued to all cadets beginning in 1975, and preconfigured computers were issued to all cadets beginning in 1986. In recent years, the department has established a UNIX workstation lab, an NSF-funded PC lab, and has run experimental sections with notebook computers and with multimedia [1,2,3, 5,11].

In the spring of 1933 West Point entered an interesting competition in mathematics against Harvard. Army was the home team, so, the Harvard competitors traveled by train to West Point. All the competitors (12 sophomores per team) took a test written by the president of the Mathematics Association of America. The West Point "mathletes" defeated Harvard in the competition that was the precursor to the national Putnam Competition [7]. Since its inception in 1984, the Academy has entered two three-person teams in the International Mathematics Competition in Modeling. USMA won the top rating in 1988 and 1993.

Under the current leadership and guidance of Frank R. Giordano, the department of mathematical sciences at the United States Military Academy has become recognized as one of more progressive mathematics programs in the country. The department has developed a strong "7 into 4" program that is exciting as well as innovative.

Core Mathematics at USMA

This paper evolves from a document that is designed to inform faculty in the Math/Science/Engineering (MSE) departments (and interested others) about the core math program. They use this document to learn what mathematical skills and concepts they can expect from their students as they progress from admission to the end of the core math sequence. They will see some of the philosophy and growth goals that we have adopted in order to help develop a diverse group of high school graduates into college juniors who are prepared to succeed in an engineering stem. They read here about our Interdisciplinary Lively Applications (ILAP) and Liaison Professor programs, which aim to achieve a more integrated student MSE experience by promoting coordination and collaboration between the Department of Mathematical Sciences and the other MSE departments. They also hear the details of our program for identifying and reinforcing required mathematical skills for entering cadets.

For the special interest of our MSE instructors, we have included in our document a detailed summary of course objectives for each core math course for the academic year. The full details are not included in this article, although a summarized version is provided for the interested reader. The principal focus of our educational philosophy is that the student thoroughly understand the concepts represented by these course objectives. These summaries of course objectives allow each MSE instructor to know what mathematical concepts the students have studied and when they studied them. As part of our educational philosophy, we recognize that mastery of conceptual knowledge is an inefficient process requiring periodic review, practice, and consolidation. Therefore, we recommend that MSE courses which rely heavily on portions of this conceptual material identify those portions to the student at the beginning of the course, and then reinforce student understanding as appropriate.

Also of special interest is the list of Mathematical Recall Knowledge - this is a modest list of basic facts that the MSE Committee has decided should be memorized by each student. The types of facts included are those that often need to be recalled quickly and without much thought in a variety of MSE courses in order to efficiently continue work on the problem at hand. These facts are added to the students' repertoire gradually as they move through the core math program - courses and lesson numbers are annotated beside each. Within the core math program, recall knowledge is not only used in each course, but the currently accumulated set is tested periodically. To make this truly recall for the students requires reinforcement by each department. For MSE courses that rely heavily on some subset of these, we recommend that they identify these to the students at the beginning of the course, and then reinforce them as appropriate. We update this document annually and provide it to all our servicing departments.

Educational Philosophy, the Role of the Core Mathematics Education at USMA

> The mind is not a vessel to be filled but a flame to be kindled.
>
> Plutarch

Core mathematics education at USMA includes both the acquisition of a body of knowledge and the development of thought processes judged fundamental to a student's understanding of basic ideas in mathematics, science, and engineering. Equally important, this education process in mathematics affords opportunities for students to progress in their development as lifelong learners who are able to formulate intelligent questions and research answers independently and interactively.

At the mechanical level, the core mathematics program minimizes the memorization of a disjoint set of facts. The emphasis of the program is at the conceptual level, where the goal is for students to internalize the unifying framework of mathematical concepts. To enhance conceptual understanding of course objectives, major concepts are interpreted both numerically and graphically to reinforce the symbolic presentations. This visceral understanding facilitates the presentation of concepts in downstream science and engineering courses.

Concepts are constantly applied to representative problems from science, engineering, and the social sciences. These applications develop student experience in modeling, and provide immediate motivation for developing a sound mathematical foundation for future studies. Additionally, these applications demonstrate future opportunities for the students and aid them in sharpening their interests.

Inherently, progressive student development dictates that education is an inefficient process and that time must be provided for experimentation, discovery, and reflection by the student. Viewed from this perspective, the core mathematics experience at the Academy is not a terminal process wherein a requisite subset of mathematics knowledge is mastered. Rather, it is a vital step in an education process that enables the student to acquire more sophisticated knowledge more independently. Within this setting, review, practice, reinforcement, and consolidation of mathematical skills and concepts are both necessary and appropriate, both within the core math program and in later science and engineering courses.

The student who successfully completes the Academy's core mathematics program should have a firm grasp of the fundamental thought processes underlying both discrete and continuous mathematics, both linear and nonlinear mathematics, and both deterministic and stochastic mathematics. The student should also possess a curious and experimental disposition, and possess the scholarship to formulate intelligent questions, seek appropriate references, and independently and interactively research answers.

Student Growth Goals: Transforming High School Graduates into College Juniors

> The greatest good you can do for [students] is not just to share your riches but to reveal to [them their] own. Benjamin Disraeli

When students enter the core math sequence, they are only a couple of months past their high school graduation. They represent a variety of backgrounds, levels of preparation, and attitudes toward problem-solving. At the end of the core math sequence, these same students will have chosen an upper-division engineering sequence and a field of study (or major), will be halfway through the requirements for a BS degree, and will be required to aggressively and successfully tackle the more synthetic and open-ended challenges of their engineering science and engineering design courses. One of the responsibilities of the core math program is to move this diverse group of entering students from the former state to the latter. Our vehicle for doing this is the set of student growth goals outlined below.

We feel that success in the MSE program at the Academy requires that, over a period of time, each entering student must mature in attitude toward the nature of problem-solving and mathematics, that each must develop and grow confidence in the processes that allow them to successfully learn and use mathematics, and that each must develop facility in the skills and arts that allow them to apply mathematics and to collaborate with others. The four core math courses are carefully coordinated so that the student can grow in each of the following areas. The result is a student prepared to perform the synthesis and meet the open-ended challenges required by their chosen engineering stem. These student growth goals are [9,10,11,13,15,16,17]:

ATTITUDES:

Mathematics is a Language

- Possesses numerical, graphical, and symbolic aspects
- Adds structure to ideas
- Facilitates synthesis

Mathematics is Thinking

- Is logical
- Requires only a few principles to be internalized

Learning Mathematics is an individual responsibility

- Requires effort
- Requires time
- Requires interaction with others

Learning Mathematics requires a curious mind

- Requires the motivation to learn

Learning Mathematics requires an experimental disposition

- Requires the ability to recognize pattern
- Requires the ability to conjecture
- Requires the ability to reason by analogy

PROCESSES:

- Learn to be confident and aggressive problem-solvers
- Learn to think mathematically (discrete, continuous, deterministic, stochastic)
- Learn good scholarly habits toward progressive student independence
- Learn to apply math to the real world

SKILLS/ARTS:

- Communicate mathematics orally and in writing
- Learn modeling as the art of applying mathematics to the real world
- Use the computer and calculator as tools for learning and problem-solving

Mathematical Thread Objectives: A Framework for Student Growth

One of the ways in which we weave together the four core math courses to attain our student growth goals is through the use of Mathematical Threads. These threads allow us to operationalize and achieve the student growth goals by translating them into specific, measurable course objectives. We have established thread objectives in six areas: Mathematical Reasoning, Mathematical Modeling, Scientific Computing, Writing in Mathematics, History of Mathematics, and Connectivity. Below are listed the goals for the core math program in each of these threads. Each core course builds upon these threads in a progressive and integrated fashion; thread objectives for each core course are summarized. These threads are [9,10,11,13,14,15,16,17]:

MATHEMATICAL REASONING:

- Interrelate numerical, graphical, and symbolic representations
- Synthesize
- Conjecture
- Apply logic
- Understand the limiting process
- Infer in situations of uncertainty

MATHEMATICAL MODELING:

- Recognize when a quantitative model may be useful
- Apply the modeling process
- Identify assumptions in a model
- Test the conclusions from a model for sensitivity

SCIENTIFIC COMPUTING:

- Appreciate the role of machines as aids in learning and doing
- Manipulate and analyze data
- Interrelate symbolic, numerical, and graphical representations

- Recognize the capabilities and limitations of computational aids
- Perform simple programming

WRITING IN MATHEMATICS:

- Communicate effectively
- Express ideas clearly and logically
- Model English with mathematics; interpret mathematics into English

HISTORY OF MATHEMATICS:

- Appreciate mathematics as a continuous human endeavor
- Motivate the further study of math
- Know key chronology and personalities involved
- Understand the service role of mathematics

CONNECTIVITY

- Understand how mathematics is related across the core curriculum
- Understand how to use the mathematical principles and ideas in other courses.

Principles and Expectations for Learning: Students and Instructors

TEACHING PRINCIPLES:

The following principles guide our entire core math program:

- Small section sizes (15 to 18 students)
- Interactive teaching
- Active classroom participation by students
- Daily preparation by each student
- Frequent feedback to each student
- Availability of additional instruction by the instructor
- High performance required of each student in class

INSTRUCTOR EXPECTATIONS OF STUDENTS:

As students progress through the core mathematics program, they grow both mathematically and academically. To help the student grow in this manner, we hold the following expectations of each student:

Student Responsibility for Learning:

- Have a foundation in the basics
- Come prepared to class
- Put forth a scholarly effort
- Work to your potential
- Participate in the instruction, discussion
- Seek assistance when needed
- Learn something about mathematics

Student Growth and Development:

- Time management
- Organization - to include course portfolios
- Meet suspenses of turn-in requirements
- Scholastic - to include spell-checking of submissions
- Proper documentation
- Reasoning/thought process
- Communication - to include checking e-mail
- Computer usage
- Standard of conduct (military courtesy, lectures, etc.)

STUDENT EXPECTATIONS OF INSTRUCTORS:

The instructor certainly has a role in this student growth. Here is what each student can expect from his or her core mathematics instructor:

- Facilitation of discussion and learning
- Challenging problems
- Applications that demonstrate the relevance of the student's mathematical education
- Activities that allow the student to participate in learning
- Timely feedback
- Fairness in grading
- Assistance/help
- Enforcement of standards

Characteristics of Core Mathematics Courses

We present core mathematics so that the student sees mathematics in terms of the following three approaches: Analytical, Graphical, and Numerical. Whenever possible they will see a problem in terms of all three models. We expect the instructor to facilitate mathematics through Discovery and Experimentation by the student, not lecture. The instructor motivates a well-integrated program by using lots of Applications, the Computer Thread, and the Modeling Thread [11]. The curriculum is designed to promote student growth.

We expect our instructors to be Interactive Teachers. They must try to understand the learning process and how to get students involved. We provide no template but we do provide extensive Faculty Development Workshops prior to an instructor teaching one of our core mathematics courses. As instructors they must recognize the difference between Education vs Training, Processes vs Algorithms, and Skills vs Rote Procedures. We develop the instructors to *Motivate Mathematics*. We provide methods and tools to aid them as motivators.

Methods:

- Analysis
- Historical Development/Problems
- Problem Solving
- Real World Situations
  - Modeling
  - Applications

Tools:

- Enhancements:
  - Computer Demonstrations/Applications
  - Special Projects
  - Case Studies
- Graphics/Visual Displays/Films

Instructor Attitudes

The following observations typify the ideas that each core mathematics instructor is asked to reflect on as he or she sets the tone in the classroom and leads the students through the core math program.

- Be yourself - be enthusiastic
- Students remember their best and worst instructors . . . be remembered for the right reason.
- Mathematics is fun - don't keep it a secret.
- A teacher affects eternity; he can never tell where his influence stops. – Henry Adams
- To teach is to learn twice. – Joseph Joubert
- Mathematics is not a spectator sport.

Mathematics Portfolios

Each semester in each core mathematics course, students are required to produce a portfolio of their course work from that semester [11,16]. The purpose of these mathematics portfolios is two-fold:

- Each portfolio is the student's "one-source" study guide for the term end exam for that semester.
- The student's accumulated set of portfolios is a mathematical reference which he or she will build upon throughout the core mathematics program, in future MSE courses, and in the core engineering stem.

In order to help achieve these objectives, portfolios are prescribed to contain at least the following:

- Title Page.
- Table of Contents.
- Gateway Exams/Fundamental Skills Exams.
- Projects.
- Written Partial Reviews (hour-long exams).
- Other items (reflective summary, drill problems, quizzes, writing assignments, etc.), as determined by the Program Director.
- Other items, clearly marked and neatly organized, that the student would like to maintain.

Students are required to maintain all of their mathematics portfolios throughout their academic careers at the Academy. Therefore, downstream mathematics, science, and engineering (MSE) departments are encouraged to make reference to and build on this material contained in students' mathematics portfolios. If students develop mathematical amnesia, they can require them to bring in their appropriate mathematics portfolios, require them to show and brief on how they used the idea at issue in core mathematics, require the students to rework and resubmit problems from their mathematics portfolios that are similar to the idea at issue. These portfolios are collected and evaluated by the instructor as a small part of the student's current course average.

Students find the collected set of mathematics portfolios to be a unique documentary of their intellectual growth, as well as a valuable summary of their mathematical repertoire. In addition, we believe that the downstream MSE instructor can use these mathematics portfolios to help improve student retention and performance of mathematics skills in their program.

The Math Clinic

The Math Clinic is a significant resource for students. It has been designed to provide a place where students can go to join other students to study mathematics. For many students, group study is a valuable form of collaborative learning.

The Math Clinic has been resourced with equipment and materials to assist students in their study. Some of these resources include:

- Carrels for individual study, and tables and blackboards for group study.
- All departmental library holdings (including over 5500 volumes and numerous periodicals) located adjacent to the Clinic, 20 feet away from the student work area.
- A reserved book section in the departmental library holdings.
- Four (4) new 486 computers currently loaded with Quattro Pro, Derive, Minitab, Word for Windows, and other cadet software.
- A large screen computer monitor (occasionally borrowed by instructors for class use).
- Course solution books (for courses that feel a need to maintain such a book).
- Written Partial Review (Exams) solutions posted for some finite period of time after the exam.
- An "On-Duty" math instructor, just waiting for questions to answer (manned from 0715-1130 and 1230-1630 weekdays).

The Math Clinic is not intended to replace the usual Additional Instruction (AI) between students and their instructors. Rather, as part of the student growth model, we expect students to take increasingly more responsibility for their own learning, and we provide the Math Clinic as a daily study resource to facilitate this growth.

The Math Clinic is also available to cadets with mathematical difficulties in other departments. An instructor in another MSE department who has a student with a mathematical problem (either remedial or advanced) can refer that student to the Math Clinic for assistance. This referral process often works best if the instructor informs the appropriate liaison professor about the referral to discuss the nature of the need and its resolution.

Interdisciplinary Lively Applications Projects (ILAPs)

"STUDENTS LEARN MATHEMATICS BY DOING MATHEMATICS"

This has long been a slogan in our department. From Crozet's time, this has meant daily student work at the blackboard which enables them to practice and explain their problem solving skills and their instructors to gauge the student's understanding. While this practice continues, reduction in class time and the advent of calculators, computers, and symbolic manipulator software now provide the opportunity for students and teachers to create and explore more sophisticated problems. In developing these 'more sophisticated problems', mathematics instructors here, as elsewhere, have naturally turned to mathematics applications in nature, science, engineering, economics, and even physical education. At the same time, teachers in these disciplines who were obtaining and using these new tools were concerned about the development of the students' abilities in their use. Thus, mathematics faculty are often engaged in conversations as to how best to prepare students to wield these new weapons of discovery and learning. In 1992, we formalized this process in an ILAP program [10,11,12,14]. Its goals are to:

- excite students with the power of mathematics as a tool in describing, investigating, and

  solving problems.
- allow students to realize their new subjects that become assessable to them as they acquire

  mastery of more mathematics.
- create a setting in which downstream departments and their future students can interact, adding credibility to the applications being studied by allowing these departments to demonstrate the utility of mathematics within their disciplines

- enhance inter-departmental cooperation and make the four year student experience more cohesive.

The ILAP program operates under the supervision of the Academy's Math, Science, and Engineering (MSE) committee. Each year, the MSE committee designates downstream ('application') departments and particular mathematics core courses to coordinate project development. Faculty from the two departments then formulate a suitable project in an applied area of interest in which a particular mathematical concept being studied in the core course is used.

During a semester, each core mathematics course assigns two or three projects that demonstrate realistic applications of the mathematics being studied. Ideally, as the project is assigned, the application department presents an introductory lecture (or video) explaining the problem situation. Then, as the project is due, the application department returns to present a concluding lecture (or video) reviewing the formulation and problem solving process, highlighting the important concepts being used and illuminating extensions of the problem in their discipline. ILAPs conducted during the past two years have included the following:

Discrete Dynamical Systems:

- 1-D Heat Transfer in a Bar - Civil Mechanical Engineering (Thermodynamics)
- Pollution Levels in Lake Shasta - Environmental Engineering
- Tank Battle Direct Fire Simulation - Systems Engineering
- Pollution Levels in the Great Lakes - Environmental Engineering
- Smog in the LA Basin - Chemistry
- Car Financing - Economics

Calculus I:

- Aircraft Ranges under Various Flight Strategies - Aero Engineering
- From Bungee Cords to the Trebuchet - Mechanical Engineering (Vibrations)
- Oxygen Consumption and Lactic Acid Production - Physical Ed.
- From Wing Resonance to Basketball Rims - Mechanical Engineering (Vibrations)

Calculus II:

- Cobb-Douglas Production Models - Economics
- Pollution of the Great Lakes -Environmental Engineering

Probability and Statistics:

- Pollution in the Great Lakes-Environmental Engineering

The creation of the ILAP flows naturally into various major service courses, such as Engineering Mathematics, which are required for those students pursuing subjects in which more advanced mathematical skills are needed.

The ILAP program embodies the belief that early mastery of mathematical skills produces in students the realization that they can indeed formulate and analyze interesting problems arising in engineering, science, business, and many other fields. Our experience is that this realization increases student motivation. We expect the application departments will see the benefits of this motivation as our students move from core mathematics into their disciplines. Our ILAP material is public domain. If the reader is interested in any of our materials, please contact our department chairman.

Liaison Professors

"SERVICING WHAT WE SELL"

Most departments at the Academy rely on at least some portion of the core mathematics program to prepare cadets for studies in their department. Additionally, the Department of Mathematical Sciences has found that the study of mathematical principles and methods is almost always made easier by considering problems in applied settings with realistic scenarios. In order to facilitate success in both of these areas, the Department of Mathematical Sciences has instituted a liaison program in which a tenured faculty member is designated as the Liaison Professor primarily responsible for coordination with a particular client department.

The major focus of this program is to achieve a more integrated student MSE experience by promoting coordination and collaboration between the Department of Mathematical Sciences and the other MSE departments. The role of the Liaison Professor is to serve as the principal point of contact for members of his client department; The Liaison Professor fields questions, accepts suggestions, and works issues from the client department with respect to course material, procedures, timing, and any other matters of mutual interest. Additionally, the Liaison Professor is a first point of contact for course directors in the Department of Mathematical Sciences who are looking for examples and applications appropriate to their course and need referrals.

This program is not intended to preclude closer coordination (for instance, at the course director or instructor level), but rather is intended to provide a continuing source of information and input at the senior faculty level which insures that interdepartmental cooperation is accomplished across several courses in a consistent fashion, as required.

Faculty Growth Model

We have developed a faculty growth model for our instructors [18]. A majority of our instructors are new master's or Ph.D.'s who are only with us about three years. Our mission is to provide them the opportunity to grow as teachers and researchers. We have developed five faculty development workshops that are attended by the faculty scheduled to teach the course associated with the workshop. These workshops vary in length from Faculty Development Workshop (FDW) I which lasts six

weeks to other FDWs (II-V) that last a few days to one week. The FDW provides course policies, strategic course overviews, explains "how" that course fits into the growth model, course technology, and how it begins to prepare the instructor for teaching the course. Instructors are involved with research opportunities in applied mathematics or operations research during one of their summers.

Scope

Our faculty consists of 61 professors, associate professors, assistant professor, and instructors. Our primary focus is the rotating military instructor (currently approximately 45 of our 61). The individuals have newly acquired masters degrees in applied mathematics or operations research. Their instructor load is 3 sections of a 4.5 credit hour course or 4 sections of a 4 credit hour course. The section sizes in our core courses are small, about 16-19 students in each section.

Within our 25% civilianization goal, we currently have 6 assistant professors at the Ph.D. level. These assistant professors are involved in our Center for Faculty Development under Professor Don Small to improve their awareness and teaching effectiveness. They are light loaded by one section in order to perform research and publish. They primarily teach in our core courses.

Our associate and full professors are our core course leaders and mathematics electives leaders. Each of the mathematics core courses is under the supervision of one of our Academy (associate or full) professors. In addition to the daily operation of the core course, they are responsible for the training of their faculty, the fairness and objectivity of the core course, the projects and testing instruments as evaluation and assessment tools, and the end-of-course surveys.

Our mathematics electives are taught by our more senior faculty. The class size is usually a little smaller, from 5–15 students. All our electives are 3 credit hours. The typical load is 2-3 classes. Electives are taught to our mathematics majors (about 20-30 a year), our operation's research majors (10-15 a year), and to our servicing departments: science, engineering, and computer sciences (over 300 a year).

Core Mathematics Summaries

Our first course is Discrete Dynamical Systems (DDS). In that course we examine discrete models through Systems as well as introduce calculus. Major topical coverage includes:

- Solving general DDS's through iteration.
- Solving 1st and 2d order, linear, constant coefficient, homogeneous and nonhomogeneous DDS's.
- Analyzing the long-term behavior of a DDS system.
- Modeling problems involving populations, decay, interest rates, and missile inventories.
- Studying linear algebra through matrix multiplication, determinants by cofactors, and eigenvalues and eigenvectors.
- Solving 2x2 and 3x3 systems of equations via Gaussian elimination and by inverting the coefficient matrix.
- Solving a system of linear, 1st order, homogeneous DDS's via its eigenvalues and eigenvectors.
- Analyzing the stability of a system of DDS's.

- Modeling problems involving voting trends, market share trends, predator-prey, mixing, decay, and the discretization of the 1-D heat equation.
- Studying sequences and their limits.
- Studying functions, continuity, and limits.
- Introducing the derivative as a limit, as an instantaneous rate of change, and as the slope of a curve at a point.
- Modeling motion problems and related-rates problems.

Our second course is Calculus I. It covers differential, integral calculus, and differential equations. The topical coverage is:

- Review limits, continuity, and derivatives.
- Model problems involving motion, related rates, and optimization.
- Analyze properties of functions using the 1st and 2d derivatives.
- Understand and be able to apply the Mean Value Theorem.
- Recognize the definite integral as a limit, and be able to approximate via numerical integration.
- Understand the Fundamental Theorem of Calculus.
- Model problems involving motion and area.
- Be acquainted with some common integration techniques (u substitution, by parts).
- Be acquainted with graphical (direction fields) and numerical (Euler's method) techniques for analyzing DE's.
- Solve 1st order DE's via separation of variables or integrating factor.
- Solve linear, constant coefficient, homogeneous DE's via the characteristic equation.
- Solve linear, constant coefficient, nonhomogeneous DE's via annihilators or undetermined coefficients.
- Review linear algebra through eigenvalues and eigenvectors.
- Solve systems of linear, constant coefficient, homogeneous DE's.
- Model problems involving growth/decay, motion, spring-mass systems, heating/cooling, and mixing.

The third course is Calculus II. In this course we analyze problems in the continuous domain involving 2 or 3 independent variables. The coverage includes:

- Review solution procedures for second order, linear, constant coefficient, homogeneous DE's.
- Review Euler's formula (cis-theta).
- Convert complex numbers between rectangular and polar form.
- Review and model spring-mass problems.
- Develop the amplitude/frequency/ phase shift functional form of trigonometric solutions.
- Understand the direction/distance approach in a plane.
- Connect rectangular and polar coordinate representations (to include arc length).
- Understand vector representations and their algebra (to include dot products and cross products).
- Become proficient in parameterization.
- Solve intersection problems involving lines, planes, and space curves (via solving systems of linear and nonlinear equations).
- Generalize single-variable calculus to vector-valued functions.
- Visualize vector-valued functions as space curves.
- Model ballistic trajectory problems.
- Visualize surfaces, gradients, and directional derivatives.
- Develop the calculus of multivariable functions.
- Find and classify extrema.
- Model problems involving production optimization, thermodynamics in plane bodies, and moments and masses of plane bodies.

Our fourth course is Probability and Statistics. It exposes cadets to stochastic problems involving both discrete and continuous processes, and introduces them to methods for analyzing same. The topics covered include:

- Use visual techniques for representing data (histograms, dotplots, stem-and-leaf plots, boxplots).
- Understand samples vs populations.
- Describe a data set (mean, mode, median, variance, standard deviation).

- Model various shipping and transportation problems.
- Understand sample spaces and events.
- Understand the laws and axioms of probability (to include representations via calculus).
- Understand mutually exclusive, collectively exhaustive, and independent events.
- Understand and apply Venn diagrams and Tree Diagrams.
- Understand the Product rule for counting (combinations and permutations).
- Understand Bayes' Theorem and conditional probability.
- Model basic reliability, reliability of circuits, transportation problems, Markov chains.
- Define and classify random variables (discrete, continuous, and joint).
- Find and use PDF's and CDF's (uniform, binomial, Poisson, exponential, normal, t, and χ^2).
- Find probabilities of random variables (including use of standard normal tables).
- Find expected values and variances of random variables.
- Understand covariance of joint random variables.
- Understand and apply the Central Limit Theorem.
- Model reliability of more sophisticated circuits and subsystems, Class IX requisitioning system, deployment of military equipment to the Gulf region, and other transportation problems.
- Use basic linear regression models.
- Understand and apply basic estimation techniques.
- Understand and apply both confidence intervals about the mean and variance with normality.
- Set up and perform hypothesis tests; understand level of significance and power.
- Model transportation problems and institutional research issues.

Richard West is currently evaluating and assessing changes to our core mathematics program. He plans to publish his results in the coming year. Additionally we periodically examine and update a strategic plan which spans the next five years. This plan accesses our standing in several areas and prioritizes our efforts for the coming year.

Last fall, our department hosted a "7 into 4" conference. As a part of that conference, we presented our core program as a strawman for a "7 into 4" core mathematics curriculum. The comments we received were laudatory with an occasional reservation that our program would not work well at their institution.

Our students are apprehensive about the Discrete Dynamical Systems course. A majority of our students have had calculus in high school and their average entering Math-SAT is above 640. This is a course of a different breed. It allows us to accomplish several important objectives:

- Transition of high school student to a college student
- Introduction of technology of graphing calculators (HP-48G) and computers
- Applications into modeling that are or can be "hands on"
- Transition into understanding limits
- Transition into calculus reform

Our students recognize this as a rare subject. They poke fun at the course in their student paper and in their spoofs on Academy life. The Academy has seen a raise in overall QPA since our introduction of this core program. The averages have risen in all our core mathematics courses. The failure rate has really fallen off. We feel that our standards are still the same and that the students are rising to achieve these standards. However, we are not satisfied with all phases and we are constantly revisiting our core program and looking for ways to improve the course and performance of the students.

References

[1] Arney, D. C., J. Hayes, and Bruce Robinson. 1987. Microcomputers in Mathematics at the United States Military Academy. *Proceeding of the American Society for Engineering Education*. 1486-1490.

[2] Arney, D. C and B. Robinson. 1988. Weaving a Computer Thread in Mathematics using Microcomputers. *COED Journal*. September: 38-42.

[3] Arney, D. C., L. Dewald, and R. Schumacher. 1990. Computer Technology and Its Impact on Undergraduate Mathematics Education. *Proceedings of the Conference on Technology in Collegiate Mathematics.* Reading MA: Addison-Wesley. pp. 9-102.
[4] Arney, D. C., L. Dewald, and J. Edwards. 1990. Mathematics at USMA for 1990 and Beyond. *Transactions of the Seventh Army Conference on Applied Mathematics and Computing.* 1-12.
[5] Arney, D. C. and L. Dewald. 1990. The Introduction of a Computer Algebra System at the United States Military Academy. *Computer Algebra Systems in Education Newsletter.*
[6] Arney, D. C. and Frank Giordano. 1992. Core Mathematics for the 1990s. *Proceedings of the Third Annual International Conference on Technology in Collegiate Mathematics.*
[7] Arney, D. C. 1988. Sources: resources in the History of Mathematics at the United States Military Academy. *Historia Mathematica.* Nov: 368-369.
[8] Arney. D. C. and R. West. 1994. Content teaching and Learning Using technology in Core Mathematics. *Proceeding of the Sixth Annual International Conference on Technology in Collegiate Mathematics.* Reading MA: Addison-Wesley. pp. 29-35.
[9] Fox, William P.1994. Using Discrete Dynamical Systems in Core Probability and Statistics. *ABSTRACTS of Papers Presented to the American Mathematical Society.* 15(1): 199.
[10] Fox, William P. 1995. Modeling Thread for a 7 into 4 Curriculum. *ABSTRACTS of Papers Presented to the American Mathematical Society.* 16(1): 259.
[11] Fox, William P. 1995. A 7 into 4 carry Through Problem. *PRIMUS.* 5(2): 163-177.
[12] Giordano, Frank. 1994. Undergraduate Lively Applications projects. *ABSTRACTS of Papers Presented to the American Mathematical Society.* 15(1): 185.
[13] Myers, Joseph D. 1994. Differential Equations as a Capstone for freshman Calculus. *ABSTRACTS of Papers Presented to the American Mathematical Society.* 15(1): 196. 1994.
[14] Myers, Joseph D. 1995. Calculus I: 3 into 1 as a means of Achieving 7-into-4. *ABSTRACTS of Papers Presented to the American Mathematical Society.* 16(1): 260.
[15] West, Richard. 1994. Discrete Dynamical Systems–Linking High School to College Mathematics. *ABSTRACTS of Papers Presented to the American Mathematical Society.* 15(1): 203.
[16] West, Richard. 1995. Integration of seven topics into four courses. *ABSTRACTS of Papers Presented to the American Mathematical Society.* 16(1): 263.
[17] West, Richard. 1995. "Evaluating the effects of changing a mathematics core curriculum", *ABSTRACTS of Papers Presented to the American Mathematical Society.* 16(1): 189.
[18] Luchins, Edith H. 1994. Preparation for Collegiate Mathematics teaching: The West Point Model from a Visiting Professor's Perspective. *You're the Professor What Next*, Betty Anne Case Editor, Notes Number 35. Washington: The Mathematical Association of America. pp. 96-99.

Section II
Core Courses and Their Contents

Discrete Mathematics in the Core Curriculum

Alan Tucker[5]

Discrete mathematics belongs in the lower-division core mathematics curriculum because its concepts and modes of reasoning underlie two very important new sectors of the modern quantitative world:

- Decision-science related mathematics such as operations research and game theory.
- Computer science and computer engineering.

While there are also areas of established mathematics that have always drawn on discrete methods, such as discrete probability, and other established areas that draw more on discrete methods now because of new computer-based explorations, such as group theory, their needs for discrete math can be addressed within courses devoted principally to these other areas. It is the pervasive and growing role of computer science and the new importance of decision-science mathematics that justifies adding discrete mathematics to the core curriculum in mathematics.

In a speech in the 1960's at the dedication of Weaver Hall, modern home of the Courant Institute, Warren Weaver offered a two-way characterization of mathematical phenomena in the world. This model was:

> Organized Simple Systems: differential equations
>
> Organized Complexity: routing telephone calls
>
> Stochastic Simple Systems: distribution of SAT scores
>
> Stochastic Complexity: weather prediction; brain function

Weaver argued that extant mathematics had concentrated on simple systems but that the important uses for mathematics in the future lay in complex systems. Mathematics had focused on simple systems because they were much more amenable to closed-form, theory-driven, mathematical analysis. There was unlikely to be comparably "nice" mathematical methods and theory for complex systems.

This is a reason why discrete mathematics has not been in the "mainstream" of mathematics research or education. It lacks theories that build layers of theorems upon layers of theorems, or powerful solution techniques as exist in calculus. In discrete mathematics, it is frequently the case that each problem has to be solved individually, from scratch, by enlightened mathematical brute force. While theoretically unsatisfying, pedagogically such situations have much to recommend them. There is no plugging into formulas because there are no formulas. Students must learn to develop their own insights and analysis.

It should be noted that discrete mathematics also contains topics like difference equations and logic where there is good theory and nice linkages to the modes of reasoning of more traditional mathematics. In this presentation, I am giving primary emphasis to parts of discrete mathematics that are most different from subjects like Calculus. As an aside, I should mention that emerging disciplines such as computational geometry that draw on computer science, discrete mathematics, and classical mathematics have theory and applications that are satisfying to all types of mathematicians. Computational geometry contains important problem areas such as computer vision, robot motion planning, and advanced computer-aided design (virtual reality-based approaches to designing complex products such as a jumbo jet) and draws upon methods of classical and modern geometry, graph theory, computer science, and electrical and mechanical engineering.

Discrete mathematics is in a sense a framework for dealing mathematically with a wide array of important complex systems. It lacks the satisfying theoretical scaffolding of more established mathematical fields like complex analysis. But out "in the field," discrete-mathematics problem-solving facility is very valuable. One example of this is a present I received a dozen years ago at the end of a junior-level combinatorial mathematics course I teach at Stony Brook. A student who finished the final exam a bit early left the classroom and then returned 10 minutes later with a bottle of expensive,

[5] Applied Mathematics Department, SUNY-Stony Brook, Stony Brook, NY 11794

20-year-old whiskey which he insisted on giving me. He said that he did a good amount of consulting as a computer programmer on the side and that my combinatorics courses had "saved [his backside] many times" in the past few months. At first, I was very confused because I did not teach anything directly about computer science or programming in the course. I taught students how to solve a variety of types of graph and enumeration problems. He said that he was finding critical uses for those problem-solving skills all the time and was developing much better programming solutions to clients' problems because of this course.

Discrete mathematics topics often entail situations where globally optimum solutions are not expected. Instead one faces the problem of trade-offs. In the past, mathematics has typically considered the structure of a mathematical system, even the trajectory over time of a moving object, as a (static) whole. In the spirit of computer science, discrete mathematics tends to decompose problems and systems into parts, either in series–multi-state processes over time—or in parallel—subparts each analyzed separately. In this sense, discrete mathematics shares some of the mindset of engineering, where complex machines and structures are constructed from diverse parts.

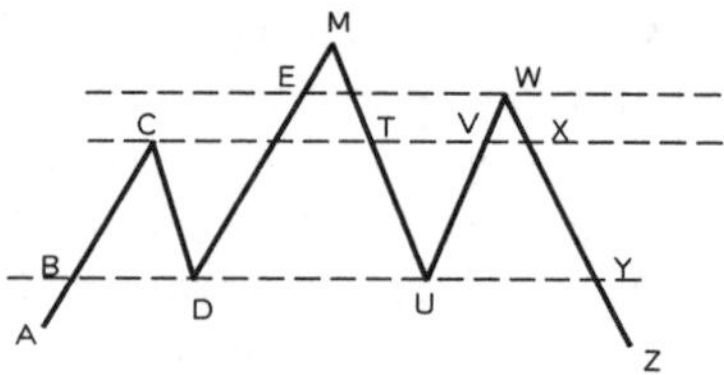

Figure 1

Educationally, the modes of thinking of discrete mathematics for organizing and analyzing information are becoming increasingly essential skills in the modern computer-driven world, whether mathematics courses address these needs or not. At Stony Brook, the first course for Computer Science majors and other students interested in a serious introduction to computer science involves no computer programming assignments. The course is devoted to problem-solving modes of reasoning. Students learn basic discrete mathematics reasoning along with styles of analysis in program development–functional programming, logical programming, and recursive programming. The junior-level combinatorial mathematics course in the Applied Mathematics Department builds on the foundation in discrete mathematics problem-solving developed in the first computer science course.

To be more concrete, here are two examples of discrete mathematics problem-solving that illustrate the preceding discussion. The first involves insights, the second involves *ad hoc* counting methods.

Mountain-Climbing

Two people start at locations A and Z at the same elevation on opposite sides of a mountain range whose summit is labeled M (see Figure 1). We pose the following puzzle: Is it possible for the people to move along the range in Figure 1 to meet at M in a fashion so that they are always at the same altitude every moment? We shall show this is possible for any mountain range like Figure 1. The one assumption we make is that there is no point lower than A (or Z) and no point higher than M.

We make a *range graph* whose vertices are pairs of points (P_L, P_R) at the same altitude with P_L on the left side of the summit and P_R on the right side, such that one of the two points is a local peak or valley (the other point might also be a peak or valley). The vertices for the range in Figure 1 are shown in the graph in Figure 2. We make an edge joining vertices (P_L, P_R) and (P'_L, P'_R) if the two people can move constantly in the same direction (both going up or both going down) from point P_L to point P'_L and from P_R to P'_R, respectively. Our question is now, Is there a path in the range graph from the starting vertex *(A, Z)* to the Summit vertex *(M, M)*. For the graph in Figure 2 the answer is obviously yes.

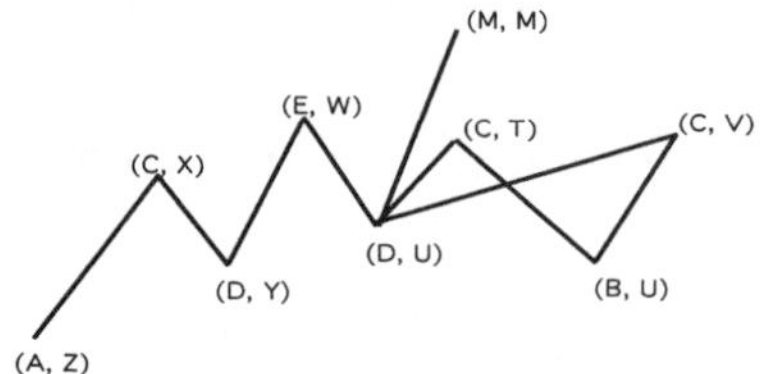

Figure 2

We claim that vertices *(A, Z)* and *(M, M)* in any range graph have degree 1 whereas every other vertex in the range graph has degree 2 or 4. *(A, Z)* has degree 1 because when both people start climbing up the range from their respective sides, they have no choice initially but to climb upwards until one arrives at a peak. In Figure 1, the first peak encountered is C on the left, and so the one edge from *(A, Z)* goes to *(C, X)*. A similar argument applies at

(M, M). Next consider a vertex (P_L, P_R) where one point is a peak and the other point is neither peak nor valley, such as *(E, W)*. From the peak we can go down in either direction: at *W*, we can go down toward *Z* or toward *U*. In either direction, the people go until one (or both) reaches a valley. At *(E, W)*, the two edges go to *(D, Y)* and *(D, U)*. So such a vertex has degree 2. A similar argument applies if one (but not both) of the points is a valley. [It is left as an exercise for the reader to show that if a vertex (P_L, P_R) consists of two peaks or two valleys, such as *(D, U)*, it will have degree 4, and a vertex consisting of a valley and a peak will have degree 0.]

Suppose there were no path from *(A, Z)* to *(M, M)* in the range graph. We use the fact that starting vertex *(A, Z)* and summit vertex *(M, M)* are the only vertices of odd degree. The part of the range graph consisting of *(A, Z)* and all the vertices that can be reached from *(A, Z)* would form a graph with just one vertex of odd degree, namely, *(A, Z)*. This contradicts the well-known fact that any graph must have an even number of vertices of odd degree, since the sum of all degrees equals twice the number of edges (each edge is counted twice when summing the degrees of all vertices). Thus any range graph must have a path from *(A, Z)* to *(M, M)*.

Shapley-Shubik Index

Consider a way of measuring the influence of different players in weighted voting. Suppose that in a 5-person regional council there are 3 representatives from small towns—call them a, b, c, who each cast one vote, and there are 2 representatives from large towns, call them D, E, who each cast two votes. With a total of 7 votes cast, it takes 4 votes (a majority of votes) in favor of legislation to enact it. Suppose that in forming a coalition to vote for some legislation, the people join the coalition in order (an arrangement of the people). The *pivotal* person in a coalition arrangement is the person whose vote brings the number of votes in the coalition up to 4. For example, in the coalition arrangement $bDcaE$, the pivotal person is c. A measure of the "power" of a person p in the council is the fraction of coalition arrangements in which p is the pivotal person. This measure of power is called the Shapley-Shubik index.

Determine the Shapley-Shubik index of person a and person D in this council, that is, determine the fraction of all coalition arrangements in which a and D, respectively, are pivotal. (By symmetry, the 1-vote people b and c will have the same index as a, and similarly E will have the same index as D.)

If a is pivotal in a coalition arrangement, then people with (exactly) 3 votes must precede a in the arrangement and people with 3 votes must follow a. Since there are only two other 1-vote people, the 3 votes preceding a must come from one 1-vote person - b or c - and one 2-vote person - D or E. Then the beginning of the coalition can be formed in 2 (choice of 1-vote person) $\times 2$ (choice of 2-vote person) $\times 2$ (whether 1-vote or 2-vote person goes first) $= 8$ ways. The remaining 1-vote and remaining 2-vote person will follow a, with 2 ways to arrange them. In total there are $8 \times 2 = 16$ coalition arrangements in which a is pivotal. There are $5! = 120$ arrangements in all, and so the Shapley-Shubik index of a is $16/120 = 4/30$.

If D is pivotal in a coalition arrangement there can be people with 2 or 3 votes preceding D. Suppose there are 2 votes before D and 3 votes after D. Either the arrangement starts with 2 of the three 1-vote people - arranged 3×2ways - then D, followed by the other 1-vote person and E in either order - 2 ways - or the arrangement starts with E, then D, followed by an arrangement of the three 1-vote people - $3!$ ways. In total, there are $3 \times 2 \times 2 + 3! = 18$ arrangements with 2 votes before D and 3 votes after D. By interchanging the people before D with the people after D in these arrangements, we obtain the arrangements with 3 votes before D and 2 votes after D. So there are 18 of the latter arrangements. In total, there are $18 + 18 = 36$ arrangements in which D is pivotal, and so D's Shapley-Shubik index is $36/120 = 9/30$.

Observe that a 2-vote person has an index 2 1/4 times the size of a 1-vote person[6].

[6] Neither of these analyses would command attention as relevant mathematical applications outside a course in discrete mathematics.

Response to Discrete Mathematics in the Core

Martha Siegel[7]

I support all that Alan Tucker said about the importance of Discrete Mathematics in the Core. Since 1986, when the report of the Committee on Discrete Mathematics in the First Two Years was released, a discrete mathematics course has become part of the curriculum at many colleges and universities—but not necessarily part of the mathematics core or even a course for the first two years.

Who takes such courses and where they are taught has influenced what discrete mathematics has become. As Alan points out, at SUNY-Stony Brook discrete mathematics is taught as a first computer science course, not in the mathematics department, and is taken by all students who take computer science. The course in the mathematics department builds on that. I am not sure we should aim for that kind of inclusion in the core—though it is one way to get 7-into-4.

The problem is that we need to get 8-into-4. At my own school, which uses a standard text, now in a 2nd edition (so we are not atypical), the course is more formal, spending quite a bit of time on logic, recursion, sets and induction with a fair emphasis on proof. Discrete mathematics *is* part of both the mathematics and the computer science core. It satisfies the faculty in neither. It is probably even less satisfying for students. It has no college mathematics prerequisite although mathematics students generally take a year of Calculus first. If we add graphs, trees, combinatorics, Boolean Algebra, etc., we have got another "course."

As we discuss having discrete mathematics as part of the core, I hope we can try to distinguish form and function, think "lively" and maybe "lean," be open to incorporation of the methods and ideas of recursion and algorithmic thinking, the spirit of cooperative work in problem solving, the core function of basic principles of logic to both mathematics and computer science and the importance of problem formulation and modeling for all our students. We can think of separate courses or work for infusion in a "Principles of Mathematics" course sequence.

My many years of experience in soliciting large projects for our Applied Mathematics Laboratory have convinced me that the majority of industrial and business problems are discrete in presentation and in solution. Our students should see genuine applications as inspiration for doing mathematics. Their appreciation of the beauty of mathematics and the joy of discovery will follow.

We should clearly articulate the goals of the core in the mathematical growth of all our students and remember that the curriculum should be designed as if most students will not take the next course, but without cramming *everything* into 4!

[7]Towson University, Towson, MD 21252

Calculus in the Core

Wayne Roberts[8]

Those with good memories will remember that calculus reform as first conceived was to create a course that was lean as well as lively. Subsequent activity has in many ways made the staid old course more lively, but as many of us have noted in our personal lives, achieving leanness is not so easy. Indeed, the more lively things become, the more difficult it is to achieve anything like leanness.

The reform project that I directed involved 26 liberal arts colleges, not one of which had a faculty that could agree on what should go into a calculus course. I entered into the project confident that we could collectively contribute a lot of good ideas, and without any hope at all of designing a single course, much less a lean course, that would meet with general approbation.

Other participants had other ideas. They took the not unreasonable position that to develop useful supporting materials, we needed to have a reasonably clear idea of what we were trying to support. Andy Sterrett came up with exactly the right principle to guide our effort to develop a core for a one year single variable calculus course.

> Never begin by asking what topics might be pruned from an existing course. Begin by asking people to identify the bare bones that must be present to support any course that you would be willing to call a calculus course.

Guided by the principle, Andy sent an open-ended questionnaire asking faculty members to make a 35 class-day list of topics that they regarded as essential to any Calculus I course, no matter where taught, when, or by whom; the same question was asked for Calculus II. A rather broadly based committee formed just for the purpose then examined these lists to compile a list of essential topics.

Besides the obvious objective of developing a list of topics, the committee had two charges. The first was that, with a semester term of 48-50 meetings in mind, their list of topics should cover no more than 35 meetings. This allows anyone teaching the course 25% of their time to convey to students the excitement they feel about favorite topics.

The second objective was to have a theme for each semester. In analogy with a play, the committee tried to think of a course in terms of a beginning that introduces the players (basic ideas), a body that develops the various sub-plots, and a conclusion that pulls everything together.

The Calculus I course was rather easy, having the fundamental theorem as a natural conclusion that does indeed draw together what has been developed regarding both derivatives and integrals. The second course was another matter, and gave rise to numerous discussions that had the committee asking itself, "What's the glue for Calculus II?" In the end, approximation won out as the theme.

This short note closes with a copy of the final report of our Syllabus Committee on a two semester course in the calculus of functions of a real variable. A shorter version, together, with an alternate suggestion that gets to multivariable calculus in Calculus II appears in Volume 1 of the *Resources in Calculus* series (Solow, 1993).

A Preliminary Syllabus for a One-year Calculus Course Prepared as Part of the Associated Colleges of the Midwest and the Great Lakes Colleges Calculus Project Funded by the National Science Foundation

Introduction

The attached syllabus has been developed first by soliciting suggestions from the faculties of the 26 participating colleges and universities, then feeding back all suggestions received to all contributors and asking them to find as much common ground as possible, and finally by forming a committee of some of the most active contributors to pull suggestions into a cohesive document. This committee has worked with the following goals in mind.

[8]Macalester College, St. Paul, MN 55105

LEAN

Our charge was as simple to state as it was difficult to achieve. Outline the basic ideas that should go into the core of any one year calculus course. The committee decided from the outset to restrict itself to ideas that could be covered in 32-35 class meetings in each of two terms. It was felt that such a syllabus would leave time for

- instructors to linger a bit over topics that they wish to emphasize
- drawing upon our resources to inject applications, cultural background, individualized projects, etc.
- testing

UNIFYING THEMES

Although most instructors indicate satisfaction with the manner in which first semester calculus hangs together with the introduction of two different limiting processes linked by the Fundamental Theorem of Calculus, many instructors indicate that they find the second semester a conglomeration of ideas and techniques, difficult to motivate. Our suggestion is to build the course around "precision and approximation," to investigate methodologies that produce exact solutions and when these approaches fail, to find ways to obtain approximate solutions with upper bounds on errors. This structure will provide the second semester with a degree of cohesiveness while simultaneously emphasizing the importance of making approximations.

ASSUMPTIONS

As we worked toward the goals above, we made certain assumptions about the manner in which the course would be taught. We list them as follows.

- Don't prove the obvious (e.g., derivative of a sum, the First Derivative Test, the Intermediate Value Theorem)
- If deficiencies are explained, less than rigorous "proofs" are acceptable (e.g., for the Chain Rule)
- Continuity should be discussed in context when the concept is needed (e.g., when discussing the Mean Value Theorem)
- Time permitting, functions defined by tables or graphs should be introduced and used in examples and exercises
- Instructors who have extra time available should expand their coverage of topics rather than introduce additional topics

Members of the committee will welcome any comments on this preliminary draft and suggestions for its improvement. We will also be happy to pass along to one of our five Working Groups your favorite ideas for making the calculus sequence lively as well as lean.

Jean Calloway (Kalamazoo College)
Bonnie Gold (Wabash College)
Harold Hanes (Earlham College)
Paul Humke (St. Olaf College)
Andrew Sterrett (Denison University), Chair
December 31, 1989

Calculus I

The intent of this syllabus is to concentrate on ideas rather than on manipulations that are more conveniently carried out with the aid of a computer algebra system or hand-held calculator. The 32 classes specified are intended to provide sufficient time in which the central ideas of first semester calculus may be discussed. Extra time that is available may be used for discussing and assigning applications (which might well require the use of a computer algebra system (CAS), testing, or developing an idea even more deeply than suggested in the syllabus. Possible applications are included at appropriate times, and the ACM/GLCA Collections are expected to provide many additional suggestions for enriching the course.

1. INTRODUCTION (1)

Describe the geometric interpretation of the two classes of problems that dominate the study of first-year calculus. Begin by describing a special case of the area problem (e.g., find the area bounded by $y = t^2, t = x$, and the t-axis) and find the limit (intuitively, of course) of upper sums that approximate the area. Then describe the tangent problem (find the slope of the tangent line to $y = \frac{x^3}{3}$ at $x = t$) and find the limit of an appropriate difference quotient. Complete the introduction by commenting on the relationship between these two apparently unrelated processes, and bring Newton and Leibniz into the picture. These two problems are easily finished in a single period with the aid of a CAS.

2. FUNCTIONS & GRAPHS (4)
   - definition, domain, range, linear, quadratic
   - trigonometric functions (sine, cosine & tangent)
   - exponential & logarithmic functions
   - composite functions
   - functions described by tables or graphs

The emphasis in this part of the course should be on graphing a broader-than-usual collection of functions (including functions given by tables) so that a rich collection of examples and applications are available throughout the year. One sequence of assignments might require students to graph a large number of functions by any method (including a CAS), with examples of each type covered to date included each day.

Note: Limits are not introduced with the assumption that the concept will be useful at a later time. Rather, the concept of limit is introduced, either intuitively or formally, when the lack of the idea would prevent one from defining and finding *instantaneous* rates of change. The notion of continuity also is delayed until it arises in a natural setting.

3. THE DERIVATIVE (10)
   - average rates of change
   - instantaneous rates of change – intuitive
   - a study of limits – intuitive or epsilon-delta
   - definition of the derivative; properties
   - derivative of polynomials
   - derivative of sine, cosine
   - derivative of exp, ln
   - derivative of sum, difference, product and quotient
   - chain rule; inverse functions

Note: In the first class, a number of applications where one computes an average rate of change should be discussed, e.g., average velocity, average revenue, average cost of sending a boxcar x miles, slope of a secant line, average rate at which a person's body assimilates and uses calcium, average rate of production at a plant, etc.

Possible Applications: motion on a line, freely falling bodies, related rates

4. EXTREME VALUES (8)
   - extreme values; rough, graphical solutions using a CAS.
   - Max-Min Existence Theorem: if f is continuous on $[a, b]$, then f attains both a maximum and a minimum value there.
   - Critical Point Theorem: Let f be defined on an open interval I containing the point c. If $f(c)$ is an extreme value, then c must be a critical point.
   - Monotonicity Theorem: $f'(x) > 0$ implies f is increasing
   - Concavity Theorem: $f''(x) > 0$ implies f is concave up
   - First Derivative Test for local extrema
   - Second Derivative Test for local extrema
   - Mean Value Theorem

Possible Applications: exact solutions of earlier word problems; obtain more detailed information of graphs of functions.

Note 1: In the first class, a number of interesting "word problems" whose solutions involve finding an extreme value should be introduced. Determine approximate solutions with the aid of a CAS.

Note 2: Use the examples and exercises from the first section to motivate the ideas that follow.

Note 3: The Mean Value Theorem presents an excellent opportunity to discuss existence theorems and how this particular one leads to several important results such as the Monotonicity Theorem and the fact that two functions with the same derivative differ by at most a constant. Finding a value for "c" should be de-emphasized.

5. ANTIDERIVATIVES & ODE'S (3)
   - antiderivatives and some basic properties
   - an introduction to differential equations: separation of variables; initial conditions

Possible Applications : exponential growth, escape velocity, freely falling bodies, Torricelli's Law

Note: Section 2 provides a large number of antidifferentiation formulas.

6. THE DEFINITE INTEGRAL (6)
   - Riemann sums
   - limit of Riemann sum
   - Integrability theorem; properties

- Fundamental Theorem of Calculus
- derivative of integral with respect to upper limit $\left(D_x\left[\int_1^x t^2\,dt\right]\right)$ in two ways

Possible Applications: areas, volumes, work, moments, distance

Note : Consistent with the introduction to the derivative, the introduction to the definite integral should begin with a number of examples where $f(x)$ is "piecewise constant." These examples are intended to illustrate that one's interest in a definite integral, and ability to recognize new applications when they occur, often begins with a function that is constant over a certain interval. For example, students know that distance equals the product of velocity and time when the velocity is constant, but students are not aware that they must define a meaning for distance when velocity varies with time.

This section provides an opportunity to remind students of the introductory lecture in which students were made aware of two important geometric problems and of the promise to show a relationship between these two problems through the fundamental theorem of calculus.

Calculus II: Exact and Approximate Representation of Numbers and Functions

The theme for Calculus II is the representation of numbers and functions by several different methods, both exact and approximate. We shall consider sequences as functions, improper integrals as limits of sequences of numbers, functions approximated by polynomials or represented as power series, functions described by the behavior of their derivatives, exact and approximate solutions to certain differential equations, and areas and volumes represented as integrals. The availability of new technology facilitates many of the techniques used to find both exact and approximate representations. We hope that teachers will keep in mind the central theme of representing a function in each of these sections and that constant interplay of precise and approximate representations will be the focus of this course.

1. INTRODUCTION (1)

Introduce several examples that illustrate the role that approximate as well as exact solutions have in mathematics. For example,

a) Remind students how to find $\sin 30^0$ and indicate how to approximate $\sin 31^0$ with a Taylor polynomial. The use of a CAS to graph the sine function and a polynomial approximation is particularly effective.

b) Indicate the integral form for finding arc length (to be derived later); discuss its limitations and how they might be overcome.

c) Indicate graphically how Newton's method generates a sequence of numbers.

d) Discuss how to estimate pi and the importance of error analysis.

2. THE DEFINITE INTEGRAL REVISITED (9 DAYS)

New applications of the definite integral, or old applications that require more sophisticated methods for finding antiderivatives or that do not possess antiderivatives, should be used to motivate the new ideas and techniques that are introduced in this section. The approach used to introduce new applications should be consistent with that used earlier, i.e., begin by adding up generic pieces rather than by memorizing standard formulas for each application.

- the definite integral: exact values from the Fundamental Theorem of Calculus
- antiderivatives by substitution, including some trig substitutions, integration by parts
- the definite integral: approximate values, Riemann sums (left, right, or mid-point), and error analysis. Trapezoidal rule and error analysis

Methods of substitution and integration by parts are included here because they are important in solving differential equations and in other mathematics courses. In the long run, other antiderivatives probably will be found by using a calculator or a CAS.

Emphasis should be placed on error analysis, including the graph of an appropriate derivative, of the numerical methods used to approximate definite integrals.

Possible Applications: arc length, including parametric representation; Buffon's Needle Problem; surface area, numerical integration of tabular data; Monte Carlo methods (throwing "darts" to approximate area).

3. SEQUENCES AND SERIES OF NUMBERS (10 DAYS)

Sequences of Real Numbers. Introduce the need to study sequences of numbers by an appropriate example. For instance, the area bounded by $y = 1/x$ and the x-axis over the interval $(1, \infty)$, the amount of paint it would take to fill the infinite funnel obtained by revolving $y = 1/x$ about the x-axis, or some of the applications listed below could be used.

- infinite sequences as functions
- limit of a sequence
- recursively defined sequences
- improper integrals, including l'Hopital's rule
- asymptotic behavior of functions (limits at infinity)

Special emphasis should be placed on the rates of growth or convergence of infinite sequences of real numbers.

Possible Applications: compound interest (n times per year and continuously compounded), approximating the numbers e or π, Newton's method for finding zeros of functions, repeating decimals as representations of rational numbers.

Series of Real Numbers

- infinite series
- geometric series

- the nth Term Test for divergence
- equivalence of series and the Limit Comparison Test
- the p-series, with emphasis on the harmonic series

Emphasize the value in comparing the rates of growth of a series of positive terms with other, known series. Begin with a derivation of the formula for the sum of a finite geometric series. Discuss convergence of the partial sums and discover conditions under which an infinite geometric series converges (diverges). Define what it means for an arbitrary series to converge and give examples to illustrate what it means for a given series to be comparable to a geometric series. Introduce both the Ratio Test and the Integral Test as "formalized comparisons." Students should be able to compare the rates of convergence or divergence for p-series, geometric series, and series involving factorials. The desire to find the exact or approximate value of a series that results from an application provides motivation for the chapter. In particular, series converging to multiples of π and e should be explored. Errors should be estimated.

Possible Applications: distance traveled by a bouncing ball; the fraction of an equilateral triangle that is covered by infinitely many circles, tangent to the triangle and each other and reaching into the corners; Zeno's paradox, the "multiplier effect" in economics

4. SEQUENCES AND SERIES OF FUNCTIONS (8 DAYS)

- the Mean Value Theorem revisited and its second degree analog
- Taylor polynomials with remainder term
- graphical comparison of a function and its Taylor polynomials; the graph of the error function for a Taylor approximation
- error estimation on intervals
- Taylor series: introduction, expansion, and a dictionary of expansions (sine, cosine, e^x, the binomial theorem)
- power series: the Ratio Test revisited and domains of convergence
- algebraic manipulation, integration and differentiation (term by term)

Possible Application: Print out a small three decimal table of the sine function.

5. SERIES SOLUTIONS OF DIFFERENTIAL EQUATIONS (4 DAYS)

Separable differential equations were introduced in Calculus I. Now we use the ideas of a previous section to obtain series solutions of second order constant coefficient linear differential equations. This unit provides an excellent opportunity to introduce applications modeled by differential equations that give rise to the sine, cosine, and exponential functions.

- defining functions via differential equations:

$$y' + y = 0; y' - y = 0$$

- the solution of linear homogeneous second order differential equations with constant coefficients, via power series

Response to Calculus in the Core I: Parallels in Calculus

David S. Heckman[9]

"Everything has been said before, but since no one ever listens, we'll say it again." *Andre Gide*

Imagine if you will that you are with me at a mathematics department meeting and the new department chair is listing the changes that we need to implement .... We need to inject applications into our courses, we need more data analysis, more geometric interpretations, more matrices, more estimation and approximation, more discovery work, more depth, less breadth, more writing, more use of technology, and on and on. Sounds like what we have just heard. The immediate reaction is if we are going to do all these new things, what will be left out or de-emphasized? The answer comes ... spend less time on factoring, less simplifying radicals, less simplifying rational expressions, fewer formal proofs, do away with conics and don't prove trigonometric identities. This is actually a mathematics department meeting in many public high schools today.

I am struck by the parallel activity that is going on in high schools and colleges today. We at the secondary level are asked to face the same difficulties that are being discussed here. The direction many secondary schools are going, i.e. to produce a core course for *all* students, is similar to the core curriculum that is being addressed in the 7-into-4 proposal. Both levels are coming to the realization that in order to service students, besides typical engineering students, who need mathematics, changes at all levels are necessary.

We want our students to come out of our program so that "When they don't know what to do, they know what to do."

I agree that the changes need to be approached from the point of view of what should be the basic content for Calculus (or Algebra) rather than what should be removed. We all have our favorite topics which are hard to give up so it is better to focus on what to keep, not what to remove. We currently have at our hands the tools to help this change. Technology gives us the opportunity to change how and what we deliver to our students. We have to change from being the "sage on the stage" to being the "guide on the side." The changes being proposed mean changes both in content and pedagogy. High school teachers know more content than elementary teachers, but we can learn a great deal from elementary teachers about methodology of teaching. The same relation exists between the high school teacher and the college professor. Professors may be better acquainted with the content, while many secondary teachers have changed how they deliver the material. We all need more opportunities to talk!

Roberts indicates a possible syllabus which concentrates on ideas rather than manipulation, numerical approximations and error analysis. This focus is to help students with the concepts of calculus. I was struck by the fact that a great many of the concerns voiced in various papers can be initiated in the high school curriculum. I just had my prealgebra class evaluate Leibniz' approximation for $\pi = 4[1 - 1/3 + 1/5 - 1/7 + 1/9 - 1/11 + \ldots]$ using a spreadsheet. The students could not do the normal manipulations but could produce and copy the pattern down 1300 rows and get a feel for the idea of what a limit is.

On the first day of class in Algebra I, using the graphing calculator, we look at the family of functions $f(x) = mx$ as m increases from 1 to ∞. The limiting case is not a function but a vertical line. A seed is planted. We investigate, using a spreadsheet, the rate of change of y with respect to x when $y = 3x - 2$. We do the same thing for $y = 2x^2 - 5$. Surprise: rates of change are not always constant. We do minimal paths and lines of best fit. Which line of best fit is the best? Many of the critical questions being raised here today can be started earlier.

The first problems I do in calculus are numerical instantaneous rate of change and area under a curve. To find the rate of change of any non-linear function at, say, $x = 3$, you can find $f(3)$ and another point close by, say $f(3.1)$, and look at the difference quotient $[f(3.1) - f(3)]/(3.1 - 3)$. I have groups use $3.01, 2.99, 2.999$ and 3.001 to get closer approximations. Thus we have the derivative (in-

[9]Monmouth Academy, Monmouth, ME 04259

formal) and limits (informal). For area under the curve, we turn on the grid on the graphing calculator and start with X-scale and Y-scale at 1. Then over the interval $[a, b]$, we just count the squares (approximating partial squares) where each square has an area of 1. To reduce the error of approximation, just change the X-scale and Y-scale to .5 giving each square an area of .25. The errors in approximation become smaller. This gives experience with informal Reimann sums and again informal limits.

How we have taught for the last 25 years is not bad; it is just that we can do better. The changes which are happening today exist because we can hook a different set of students. A college professor I know claims that getting colleges to change is like moving a cemetery: there's no help from the inside.....I don't think that this is universal. We are all here, but we have our work cut out for us. The only people who see obstacles are those who take their eye off the goal. College and high school teachers have different strengths. Let's use each other. Just remember, when the winds of change blow, some people build shelters, and some people build windmills. Let's go out and build some windmills.

Response to Calculus in the Core II: Calculus and Other 7-Into-4 Issues

Stephen Rodi[10]

In responding to Wayne Roberts' paper on Calculus in the Core, I will make three kinds of comments. Some will apply to calculus directly. Some will be about other 7-into-4 issues. Some will be about how all these matters relate to and affect community colleges.

The 26 Great Lakes and Midwest colleges followed a sensible procedure in developing their version of a lean calculus. It was smart to start with a *tabula rasa* vis-a-vis calculus and build up on the blank tablet what is essential to the calculus course rather than take an existing syllabus and pare it down. Their effort has produced a syllabus as lean as anyone could expect. Exclude anything else and many would think the knife had reached too deep and was scraping the bone.

However, a difficulty remains. This calculus reform project, like most of the other projects funded by the National Science Foundation, has ended up a two-semester, one-variable *calculus* course. These courses are generally leaner, hopefully livelier, pedagogically sounder, and, hopefully, more accessible to a wider range of students. But, for the most part, they are still two-semester, single-variable courses which do little to integrate the other material (differential equations, linear algebra, finite mathematics) which make up the 7-into-4 approach. Having completed a reform calculus curriculum, students may not be much further along in trying to "cover seven courses in four" than they would have been had they finished a old-fashion standard calculus course.

Whether the intellectual suppleness students achieve in reform calculus will be enough to allow students to then complete "five into two"—what would remain after a two-semester single variable calculus course—is not yet clear. However, I doubt that will be one of fruits of calculus reform.

As I see it, there are only two ways to achieve the efficiency of 7-into-4. The first is to take a highly abstract approach, using the synthetic power of Bourbaki-like mathematics, to interrelate all the material. This would have to begin in the first semester of the four and continue high-powered for four terms. This approach is not advocated; it would not be a productive way to approach the problem. All the pedagogical movement of the last decade has been in a direction opposite to abstract efficiency. Starting in the earliest grades and continuing at least through reformed calculus, there is a renewed emphasis on manipulatives, experimentation, projects, mathematics laboratories, computer simulation, teams and groups, and other concrete, tangible ways of learning mathematics. All of this is consistent with my favorite—and now too often used—epistemological dictum from the medieval scholastic philosophers: *Nihil est in intellectu nisi prior in sensibus.* Nothing is in the intellect unless it is first in the senses. The senses are the door of the soul, opening the mind to the individual concrete experiences from which it will abstract to both find and create mathematics. (We indeed construct our own knowledge but our construction can only reflect and elucidate the mathematics already present in nature.)

This modern pedagogical movement in mathematics, with its roots in ancient understanding of how humans learn, is not going to be reversed any time soon. It makes the "abstract" solution to the 7-into-4 approach unfeasible. In fact, it makes the solution of the 7-into-4 approach harder. One must achieve the compactification of content in an instructional format inherently more time consuming.

If a move to the abstract is not an acceptable way to achieve 7-into-4, one must look to the calculus course itself, or rather the first two semesters, as the place where significant consolidation must occur. The calculus reform movement of the last decade did not have this consolidation as one of its goals. Hence, none of the currently existing calculus reform materials seems particularly suited to advance the 7-into-4 effort. None of them integrates enough linear algebra or differential equations or three dimensional topics into the first two courses to make a noticeable contribution to the consolidation. They should not be faulted for this. They did not

[10]Austin Community College, Austin, TX 78701

set out to make such a consolidation. In fact, we can ask if such a consolidation is contrary to their underlying philosophy. Such consolidation inevitably will make the first two courses more sophisticated and perhaps less manageable by the broader base of students the calculus reform movement is in part trying to reach.

These comments raise some questions about the feasibility of 7-into-4. Equally valid are questions about its desirability. On a national scale we are attempting to educate a much broader base of students in collegiate mathematics as the foundation for an increasingly technological society and work force. Is presenting more mathematics in a shorter period of time the best way to achieve this? The last time we tried such an experiment in college by moving calculus to the freshman year (well into the 1950's and even into the 1960's calculus was a sophomore course) we increased the numbers of students who could not manage their first collegiate mathematics experience and arguably inadvertently contributed to the diminished collegiate undergraduate mathematics performance we are now trying to reverse. Humans have long developmental periods—gestationally in the womb, maturationally to adulthood, and intellectually in the mastery of complicated ideas. It is not clear the human condition always is best served by trying to do more faster. Mental percolating time is important.

The other aspect of the Great Lakes/Midwestern Colleges calculus reform which caught my ear was the idea of establishing a theme around which each semester was constructed. The second semester theme of approximation and estimation seemed appropriate and useful. This is an aspect of mathematics and calculus too easily lost sight of in the perfectionist, exact result Bourbaki approach to our subject.

If you will permit a brief digression, but one related to the idea of course theme, I have always told the many part-time instructors I have oriented over the years that they should think of a course as a play with a beginning, a middle, and an end. There is an appropriate introduction and development of material, some high points, and a denouement that gives completion. Each class should be a mini-version of such a play. The danger for students is that they will think of themselves as a passive audience, nodding approval and even applauding as the play develops, but unable to step on stage as a player/actor when asked to do so, say, for example, on a test. (Students always wonder: it seemed so easy in class when I was watching you do it but I could not do it on the test. Not surprisingly, a theater goer thinks Olivier's *Hamlet* looks effortless, unless he is called on stage to try a soliloquy.) There is less danger students can remain passive in most of the alternative class structures currently gaining popularity.

The real point of mentioning course theme, however, is that this gives a clue of how to attack the 7-into-4 approach. What needs to be defined are four major foci around which four semesters of mathematical work can be developed, covering the major topics and themes we now find in the first seven courses. This is the hard curriculum development work which must follow on conferences like this one if indeed 7-into-4 is ever to become a reality.

Finally, a few comments about community colleges. On the one hand, 7-into-4 may cause few problems for community colleges. The faculty there, both full-time and part-time, who teach these courses usually teach all of them, calculus as well as differential equations and linear algebra. So I would expect that each college would have a core faculty fully capable of teaching newly organized courses which cover this material over four semesters in an integrated formats. There may be some problems in training tutors and support staff but no more severe than what occurs in facing calculus reform. (For a long discussion of the special problems and issues that are associated with calculus reform at a typical community college, see the article I co-authored with Sheldon Gordon in the volume *Preparing for a New Calculus,* Number 36 in this Notes Series.) Hence, on an intellectual level, I think community colleges by and large would have no more trouble adjusting to a 7-into-4 curriculum than four year schools.

However, community colleges are in a very precarious position when it comes to initiating such far reaching reforms unilaterally. Community college students are very pragmatic. They want to take courses which they know will transfer to baccalaureate institutions and count in degree plans when they transfer. Indeed, community colleges have an ethical responsibility to try to make sure this happens. Hence, community college participation in any 7-into-4 experiment or structure may well depend on regional cooperation between them and major state institutions which receive their students. 7-into-4 is an area in which articulation is even more critical than in calculus reform.

Calculus after High School Calculus

Don Small[11]

Introduction.

Students who have successfully completed a full year of calculus in high school present different challenges and opportunities to colleges than do those students who have not previously studied calculus. In general, colleges have responded in one of three ways to this situation. They have attempted to advance these students by placing them into the next calculus course, excused them from calculus altogether, or had them start calculus over again. Each of these choices is primarily driven by a felt need to expose students to the content of calculus. A different approach, one centered on student growth will be advocated in this paper.

In a time when the amount of factual knowledge is exploding and technology for assessing and manipulating data is rapidly expanding, our focus needs to be on developing our students' academic growth; and, in particular, their abilities to learn on their own. Students coming out of high school calculus courses offer colleges a unique opportunity for doing just this. These students usually have received a broad and procedurally oriented introduction to calculus of a single variable. Their technical knowledge exceeds both their conceptual understanding and their mathematical maturity. Our challenge is to offer a program that will lead students to take responsibility for their own learning in a way that will develop their critical thinking skills and will lead them toward conceptual understanding of the basic ideas of the calculus. In brief, the challenge is to accelerate the development of mathematical maturity within our students. Involving students in generalizing their high school calculus experiences and holding them accountable for their learning are effective avenues for developing mathematical maturity. A key to success in this endeavor is to have students reverse the traditional way of thinking of the roles of class work and homework. Students need to realize the primary importance of out-of-class work and the secondary importance of in-class work. We need to be clear that the purpose of class work is to support outside study and not the other way around.

Any model for a 7-into-4 curriculum will expect students to take on a greater responsibility than has been true in the past for learning material that is not specifically covered in a course. How and when will students develop the expertise to do this? A partial answer is to develop a special calculus course for students who, in high school, have successfully studied calculus that:

- integrates the treatment of one and several variables;
- involves students in generalizing their high school calculus;
- emphasizes the development of mathematical maturity.

(For the purposes of this paper, "successfully" is interpreted to mean the student has received a 3 or higher on an Advanced Placement exam or received a course grade of B or higher in a secondary school calculus course equivalent to an Advanced Placement course.) In this special course, the high school calculus is viewed as a special case within the context of calculus of several variables. This view and the integrated treatment, encourages a generalization approach rather than the "building on" or "extending" approach which usually characterize programs designed to advance students to the next course. The distinction between these approaches is important to note when considering the development of conceptual understanding and maturity. The process of generalizing from a special to a more general situation requires a greater understanding of the special case than when viewing the extended result as additional content. In particular, students need to question assumptions, to understand the limitations, and to recognize relationships implied by the special case. This type of "review" activity enhances the understanding of high school calculus topics and develops mathematical maturity.

The first section of this paper sets out goals for this special course. The second section discusses characteristics shared by several students who have studied calculus in high school and the third section describes important aspects of the proposed course.

I. Goals.

A. Student Growth:

[11]U.S. Military Academy, West Point, NY 10996

- Learn how to learn and how to communicate mathematics (questioning, reading, writing, speaking, thinking, modeling);
- Accept responsibility for one's own learning;
- Develop an exploratory attitude toward learning mathematics;
- Develop an approximation approach to mathematics;
- Develop conceptual understanding;
- Develop an appreciation for calculus.

B. Content:

- Generalize the concepts of nearness, differentiation, and integration to higher dimensions;
- Taylor Series;
- Differential equations through variation of parameters;
- Broad-based review of calculus of a single variable.

II. Characteristics of students who have successfully completed a full year of calculus.

Although for purposes of this paper, I am considering students who have studied calculus in high school, I believe that the following characteristics are shared by the majority of students who have taken only one year of calculus in high school or in college.

1. Strong mechanics on *routine* exercises. For example, students do well in differentiating polynomial functions, but have a great deal of difficulty in differentiating a piecewise defined function even when each piece is defined by a polynomial. Another example involves the chain rule. Given that $f(u) = u^3 - 2$ and $u(x) = x^2 - 5$, most students can compute $(f \circ u)'(2)$. However, fewer students can make this computation when told that f and u are differentiable functions and that

$$\begin{array}{rrr} u(1) = 3 & u(2) = -1, & u(3) = 0, \\ u'(1) = 5 & u'(2) = 4, & u'(3) = -2, \\ f(3) = 4 & f(-1) = 0, & f(0) = 2, \\ f'(-1) = 3 & f'(0) = 2, & f'(3) = 6 \end{array}$$

2. Familiarity with most of the calculus vocabulary. This provides both an advantage and a disadvantage. The advantage is that starting a course with a catalogue of functions is more understandable to students who are already familiar with the names, e.g., exponential function, natural log function. A disadvantage is the limited understanding of the meanings of certain terms and erroneous connections that students have established in their minds. For example, many of my AP students believe that all functions have to be continuous.

3. Broad, uneven, and shallow coverage of single variable calculus. I gave a test 5 weeks into the calculus course to students who had received a 3 or better on an Advanced Placement test. Seventy percent of the students could not give any interpretation of the difference quotient in the definition of the derivative. Another example is excellent skill in the use of integration techniques, but minimal understanding of a Riemann sum construction.

4. High "math ego" and strong loyalty to previous program. It is natural that students who have been successful as measured by high grades, have a high regard for the program in which they did well. It is also to be expected that many of these students will resist "changing the rules of the game" from emphasizing mechanics to emphasizing conceptual understanding.

5. Weak understanding of definitions and theorems. Students often do not distinguish between definitions and examples. In general, students do not have experience in analyzing a definition to determine its limitations. With respect to theorems, students generally key in on the conclusion and ignore the hypothesis. The relationship of the hypothesis to the conclusion of a theorem is often neither understood nor appreciated by students.

6. Poor understanding of the function concept. Most of my students use the terms "function," "expression," and "equation" interchangeably. Domain and range considerations are generally overlooked, as are piecewise or discrete functions. For example, the "cut and paste" approach to writing has not carried over to constructing piecewise defined functions.

7. Strong dependence on the instructor and class time. Students expect to be able to do the homework by mimicking exercises that the instructor does in class or examples given in the textbook. For many students, the instructor is the primary, if not the only, resource. Furthermore, students feel

that what is done in class defines the course, not what is done outside of class.

8. Minimal experience in learning how to learn. Although color coding of texts has reduced the need for student highlighting, it has not resulted in aiding students to distinguish relative importance among theorems, definitions, techniques, etc. Students generally do not understand, or appreciate, the development of a topic within a text. For example, students seldom ask about the purpose of a worked example or why the author(s) selected that particular example.

III. A Proposed Course: Calculus after High School Calculus

The proposed course fits easily into time periods covering from 80 to 100 lessons. A syllabus for the course is contained in the appendix. Although the content material is similar to that in a traditional three-semester calculus program, this course is distinguished by its strong focus on student growth concepts. As stated in the introduction, a key to the success is having students grow to understand that the most important component of the course is the mathematics they do outside of class and that the objective of class work is to contribute to their outside efforts. Thus it is neither expected nor necessary for the instructor to cover all content material in class. Integration techniques, surfaces and volumes of revolution, and polar coordinates are example of topics that can be assigned outside of class, but tested within class. Class time should be primarily reserved for activities that enhance conceptual understanding, "What-iffing" homework exercises, student presentations, and learning how to learn from a text. For example, time spent having students discuss the purpose of a worked example or the relative importance of the theorems in a section or the reason for an algebra theorem is time well spent.

Learning how to learn from a text can be facilitated with a Study Guide that provides a model for approaching new material. I use a Guide that outlines each lesson in the following manner:

- Primary Reading, Supplemental References, and Homework Assignment
- Essential Terminology
- Lesson Objectives
- Study Questions
- Comments

In the first part of the course, each of the categories in the Study Guide is filled out in detail. As the course progresses and the students grow in independence, less detail is provided. At the end of the course only the outline headings are presented and students are organizing their own guide for each lesson.

I will address five major components of the course and then comment on specific content issues. Since a primary goal of the course is increasing students' mathematical maturity by focusing on student growth, my comments are designed to illustrate several pedagogical aspects of the course.

1. Integrated Approach.

The integrated approach provides students with opportunities to generalize their high school calculus to several variable calculus in ways that require them to gain a deeper understanding of single variable calculus. For example, generalizing the limit concept to higher dimensions forces the student to rethink "one sided limits." Another example is the insight that is brought to the fundamental role of the linear approximation theorem through the process of generalizing differentiation to functions of several variables.

Integrating the treatment of one and several variables offers many advantages, in addition to placing the emphasis on concepts rather than dimension. In particular, the integrated approach stresses generalization (e.g., number to vector, line to plane, and function, limit, continuity, derivative, integral from R^1 to R^n). Generalization, in turn, can provide a motivation for learning. For example, a natural question to ask when generalizing a concept is: What properties are preserved under the generalization? In answering this type of question, the student gains a deeper understanding and appreciation of the concept than he/she had the previous year. My students usually assume that all properties are preserved under generalization. Thus, in the beginning of the course, they are surprised to learn that the trichotomy law is not preserved when number is generalized to higher dimensions. What happens in this case, and what is really important, is that the students were forced to think about the trichotomy law which they had not previously done. Rethinking an idea often leads students to a careful consideration of pertinent definitions and theorems and their origins rather than just examples.

The need to understand these definitions and theorems in order to generalize provides a motivation for learning.

By viewing the high school calculus as a special case within the larger calculus context, the integrated approach provides an environment that leads students to think in terms of generalization. This, in turn, helps spur an exploratory attitude. I would like a student studying numerical integration to ask: What would happen if one approximated the function with a cubic polynomial rather than a quadratic polynomial as in Simpson's Rule or a linear polynomial as in the Trapezoidal Rule or a constant polynomial as in a Riemann Sum? Although none of my students have asked this particular question, a few this semester have asked if there was a general procedure for developing numerical integration techniques and several have questioned how the coefficients were obtained in the error bounds for the Trapezoidal and Simpson's rules.

The development of multivariable topics within an integrated approach influences the treatment of the corresponding topics in one variable. For example, the linear approximation theorem plays a central role in the generalization of the derivative to functions of several variables and thus this theorem should be clearly understood at the one variable stage. In the past, however, when I taught a traditional one variable course, this theorem was usually omitted. The processes for finding extremal values provide another example. For functions of one variable, extremal values are usually found by plotting. However for functions of several variables, one needs to understand the analytical and numerical approaches. Thus these approaches are developed for functions of a single variable and then generalized to functions of several variables.

Providing a broader context for discussions is another aspect of the integrated approach. I customarily begin each class with a student giving a 5 minute presentation on a topic that I have assigned. A second student then leads a 5 minute question period on the topic. Being able to talk about an idea in different dimensions provides students with a richer experience than when they are restricted to one dimension. These short sessions are designed to deepen understanding as well as to highlight student questioning.

2. Core Approach.

The rationale underlying the Core Approach is that if students learn how to develop a small set of concepts well then they will be able to generalize their experience to other situations on their own. Thus emphasis is placed on identifying a minimal set of "building blocks" and on developing a uniform approach. The major concepts are those listed in the statement of goals: *nearness, differentiation,* and *integration.* Four categories of functions are considered: polynomials, periodic, exponential and logarithmic, and sequences. Our set of building block functions is $x^n, \sin(x), \cos(x), e^x, \ln(x)$.

The uniform approach used to develop all of the major concepts contains five steps:

- *Motivation:* Establishing an interest in understanding as well as a need to understand an idea is the most important step. This is usually approached through group projects, readings, or class activities.
- *Develop a definition:* This is done using the Basic Approximation Process:
  - Approximate the unknown quantity with a known quantity;
  - Determine a way to obtain a better approximation;
  - Generate a convergent sequence of approximations such that each approximation is better than the previous one;
  - Define the desired property to be the common limit of all possible sequences from above. When there is no common limit, the desired property is said not to exist.

  Thus the derivative is the limit of a sequence of average rates of change; the integral is the limit of a sequence of approximations; an improper integral is the limit of a sequence of proper integrals, the sum of a series is the limit of a sequence of partial sums, etc.
- *Application of the definition:* This is done in the context of the building block functions.
- *Algebra Theorem:* Show how the concept behaves with respect to the standard arithmetic and functional operations. For example:

  If $\lim f(x)$ and $\lim g(x)$ both exist, then
  1. $\lim_{x\to a}(f(x) + g(x)) = \lim_{x\to a} f(x) + \lim_{x\to a} g(x)$
  2. $\lim_{x\to a}(f(x) - g(x)) = \lim_{x\to a} f(x) - \lim_{x\to a} g(x)$
  3. $\lim_{x\to a}(f(x) \cdot g(x)) = \lim_{x\to a} f(x) \cdot \lim_{x\to a} g(x)$
  4. $\lim_{x\to a} \frac{f(x)}{g(x)} = \frac{\lim_{x\to a} f(x)}{\lim_{x\to a} g(x)}$ provided $\lim_{x\to a} g(x) \neq 0$

5. $\lim_{x\to a} cf(x) = c \lim_{x\to a} f(x)$ for c constant.

- *Applications:* Consideration of both theoretical and applied applications of the concept.

The purpose of explicitly identifying each of these steps is to call students' attention to a structural model for developing a mathematical concept. Understanding such a model is important to students learning how to learn on their own. Very few of my AP students have ever considered how a mathematical concept is developed. Furthermore, even though they have learned many of the components of an algebra theorem, these students have never thought of the role that an algebra theorem plays in the development a mathematical concept.

3. PROJECTS.

A project is a group activity that consists of solving a multi-staged problem and writing a report describing the group's efforts. The primary objective of projects is to develop students' creative reasoning and communication skills and to do this within a group setting. Modeling, data collection, data analysis, generalization, experimentation, and exploration are common activities involved in solving a project problem. The written report is an important aspect of a project. These reports contain a title page, a one or two page summary, and appendices for graphs, computations, computer printouts, and list of references used. The summary page(s) include a brief statement of the problem, description and rationale of the solution process used, results, interpretation of the results, and an explanation of why the results are reasonable.

I generally have three or four students work together as a project team and expect that they will spend 4 to 6 hours outside of class working on the project. Each team submits one report and the same project grade is assigned to each of the team members. Projects are developed for different purposes. Here are a few examples of project problems:

Motivate a concept:

- To motivate and introduce students to functions of several variables, I generally start my course with a "Stairway Comfort Function." Students are asked to develop a function having at least 3 inputs that quantifies the comfort felt when going up or down a stairway. The output is restricted to (comfortable, uncomfortable, convert to ramp, dangerous). Students are required to analyze at least 10 stairways in the process of formulating their function. An interesting insight to stairways (and students) is gained by everyone, including the instructor.
- Motion experiments (e.g., rolling a ball, driving a car, draining a bathtub) are used to develop a sequence of average rates of change leading to the definition of an exact rate of change.
- Determining the number of 5 pound bags of grass seed required to seed a given odd shaped field is used to introduce the topic of area. (One pound of grass seed covers 400 square feet.)

Develop mathematics:

- Develop a "Thrill Function" for a roller coaster. (Nice opportunity to develop an interpretation for the third derivative.)
- Develop a numerical integration technique, with error bound, that uses a cubic polynomial for the approximating polynomial.
- Improve on Euler's method.

Applications: Bungee jumping, wheel suspension, designing the shape of a roller coaster, highway design.

Library search: Discuss Cantor's method of showing a one-to-one correspondence between the integers and the rational numbers, and that there is no one-to-one correspondence between the rational numbers and the real numbers.

4. APPROXIMATION AND ERROR BOUND.

Approximation is the backbone of calculus as indicated by the fundamental role that the Basic Approximation Process plays in the development of calculus concepts. Since the sequence (of approximations) approach is used to develop limits, every concept defined in terms of a limit has a built-in approximation process. Approximations by themselves however are not enough; one needs to quantify the value of an approximation by establishing an error bound. Thus determination of error bounds is an important consideration throughout the course.

A standard exercise that pervades the course is to approximate a specified quantity to within a given accuracy (i.e., epsilon). In the first or second week, students obtain polynomial approximations of functions to within a specified accuracy by means of plotting. Later students determine a bound on h such that the difference quotient function,

$$dq(x,h) = \frac{f(x+h) - f(x)}{h}$$

approximates the derivative function of f to a specified accuracy. This leads to the Mean Value Theorem and the Linear Approximation theorem. The generalization of the latter to vector valued functions and then functions of several variables is central to the generalization of the derivative concept to functions of several variables.

The course strongly emphasizes numerical integration, with its approximations and error bounds, in contrast to closed form integration. When numerically approximating a sum (i.e., an integral) by a method for which there is no known error bound, students are faced with the question: "How do you know when you have obtained the desired accuracy?" For example, before an integral formula for arc length has been developed, how would you approximate the length of the graph of $y = \cos(x)$ over the interval $[-1, 3]$ with accuracy 0.01? This type of accuracy question leads students into developing heuristic methods. Contrasting heuristic bounds and known bounds is a fertile arena for student explorations which leads to deeper understanding of integration than would any number of weeks applied to closed form integration techniques.

5. Use of Technology.

Students are expected to use graphing calculators and computers. Whenever possible, I replace a class with a lab once a week. In the lab, students work in pairs and each pair submits a written lab report. Being able to shift and scale graphs of functions is an important skill that is used throughout the course. Thus learning how to shift and scale graphs is the mathematics that is addressed in the first lab, the "Getting Started" lab. As just discussed, approximation is the "backbone" of the course and thus the second lab is devoted to learning how to graphically obtain a "pseudo Taylor series" to approximate a function to within a given accuracy. To reinforce the emphasis on discovery and explorations, numerous lab exercises involve using the computer to generate data from which the students conjecture a pattern and then use the computer to check their conjecture. For example, find a formula for the nth derivative of the product of two functions or a formula for the integral of $e^x \sin(x)$.

Remarks on Content

The Core approach and insistence that students assume responsibility for their learning allows the emphasis to shift from covering content to opening up access to content. As a result a number of topics can be omitted or have their treatment reduced. Here are a few examples of traditional content that I omit wholly or partially:

- The need for graphing pervades the whole course including graphing rational functions, yet no class time is spent on this topic.
- Root finding algorithms (bisection method, Newton's method) can be addressed in projects without requiring class time.
- Surfaces and volumes of revolution do not need to be covered.
- One or two days is sufficient for integration techniques and this time should be spent emphasizing the transformation aspect.
- Polar, cylindrical, or spherical coordinates can be introduced through projects with suitable references, yet no class time is involved.
- One day is sufficient for convergence tests of series. Early and continued emphasis on polynomial approximations reduces the time spent on Taylor series.
- Strong emphasis and use of sequences throughout the course eliminates this part of the traditional section on sequences and series.
- Hyperbolic functions are not covered per se, yet may be involved in exercises with a reference to their definition.
- The conic sections are omitted.
- Stokes' and the Divergence theorems are usually omitted.

The Core approach requires that students develop a mastery of basic computation skills and basic understanding of concepts. To demonstrate this mastery students must pass gateway tests, at the 100% level, in graphing, differentiation, and integration. The majority of students take these tests two or three times before passing. An example of a gateway test is included in the appendix.

Conclusion

Targeting the development of mathematical maturity by focusing on student growth rather than on content is a productive and responsible way to meet the challenge presented by students who have successfully studied calculus in high school. Critical thinking skills and the ability and discipline to take charge of one's own learning are the building blocks for success in today's technological world. Developing these characteristics needs to be central in our teaching.

Viewing high school calculus as a special case within the larger context of several variable calculus, suggests an integrated approach that emphasizes generalization. This process leads students toward conceptual understanding and involvement in discovery and exploratory work. This, in turn, accelerates the growth of mathematical maturity.

Syllabus for Calculus after High School Calculus

1 Lab "Graphing"
2 Terminology
3 Functions
4 Compositions of Functions
5 Lab: "Graphical Approximations"
6 Approximations
7 Approximations and Convergence
8 Limit Concept
9 Lab: "Fitting Curves to Data Points"
10 Continuity
11 Intermediate and Extreme Value Theorems
12 Gateway Test (Graphing), Review
13 Differentiation
14 Lab: "Local Linearity"
15 Differentiation
16 Lab: "Best Linear Approximation"
17 Algebra of Derivatives (Chain Rule)
18 Mean Value & Linear Approximation Theorems
19 Lab: "Derivative - Geometric Significance"
20 Lab: "Extreme Values"
21 Review
22 Test
23 Review
24 Picture Project assigned
25 L'Hospital's Rule
26 Partial Derivatives, Picture Project due
27 Vectors
28 Lab: "Warm-up" Exercises
29 Vectors
30 Derivatives of $f : R \to R^n$
31 Directional Derivatives
32 Lab: "Interpreting Cross Products"
33 Differentiation of $f : R^n \to R$
34 Differentiation of f: $R^n \to R$
35 Geometry
36 Lab: "Chain Rule"
37 Review; Area Project assigned
38 Review, Gateway Test - differentiation
39 Lab: "Extrema in R^n"
40 Extrema in R^n
41 Lagrange Multipliers
42 Lab: " Extrema in R^n; Area Project due
43 Test
44 Area
45 Area; Equal Division Project assigned
46 Riemann Integral
47 Lab: "Numerical Integration"
48 Numerical Integration Project assigned
49 Numerical Integration
50 Derivation of Error Bound for Trapazoidal Rule
51 Properties of Integrals
52 Review
53 Lab: "Numerical Integration II"
54 Discussion of Numerical Integration Project
55 Arc Length
56 Fundamental Theorem of Calculus
57 Lab: "Analyzing an Integral"
58 Transformation Tech. of Integration
59 Improper Integrals
60 Natural Logarithm
61 Natural Exponential Function
62 Line Integrals
63 Fundamental Theorem of Line Integrals
64 Iterated integrals
65 Test, Cow Grazing Project assigned
66 Taylor Polynomials
67 Taylor's Theorem
68 Infinite Series
69 Lab: "Harmonic Series"
70 Geometric Series
71 Convergence Tests
72 Power Series
73 Lab: "Power Series"
74 Introduction to DEs
75 Modeling, Num. Approx. Project assigned
76 Direction Fields
77 Separation of Variables
78 Linear, First Order
79 Constant Coefficients
80 Higher Order DE's , Project due
81 Application: Suspension System
82 Linear, Non-Homogeneous

83 Particular Solutions
84 Application: Suspension System Again
85 Test
86 More Than One Dependent Variable
87 Systems of DEs

Gateway

Differentiation

1. Find the following derivatives:

   - $$\frac{d}{dx}(x^5 - 4x + 7)$$
   - $$\frac{d^2}{dx^2}e^{-(x^2)}$$
   - $$\frac{d}{dx}\log(\sqrt{3x+9})$$
   - $$\frac{d}{dx}\frac{x^{3/2}}{e^{2x}}$$
   - $$\frac{d^3}{dxdydx}(xe^{-y})$$

2. Let $f : R^3 \to R$ be defined by $f(x, y, z,) = \sin(xy) + \ln(z)$. Evaluate the derivative of f at $(2, 0, 1)$.
3. A curve is given parametrically by the vector valued function $g : R \to R^3, g(t) = (t^2, e^{-t}, 1/t)$. Determine the tangent vector at $(1, 1/e, 1)$.
4. Given that f and g are twice differentiable functions defined for all real numbers and:

$$\begin{array}{ll} f(2) = 7, & g(4) = 3, g'(2) = 6 \\ f'(6) = 4 & g'(4) = 4, g''(3) = 6 \\ f'(4) = 6, & g''(2) = 3, g''(4) = 3 \end{array}$$

compute the derivative of $f(g'(x))$ at the point $x = 2$.

5. Given the following graph of $y = f(x)$, sketch the graph of $y = f'(x)$.

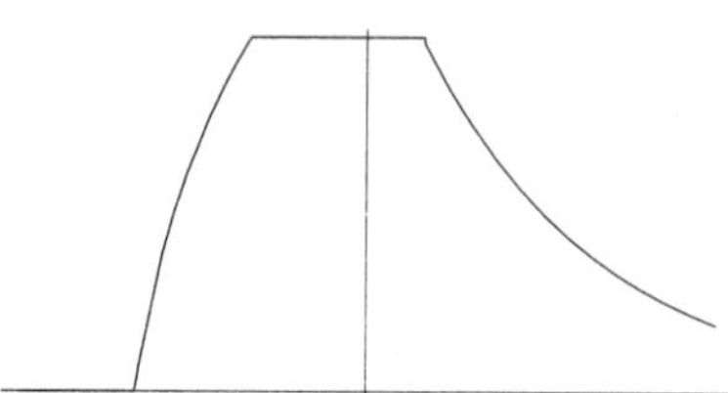

6. Let $y = f(t)$ be the number of acres of forest infested with the gypsy moth. Describe each of the following situations by assigning numerical signs (positive, negative, zero) to the first and second derivatives of f. Explain your reasoning:

   - The infestation is spreading, but at a slower rate than in the previous time period.
   - The infestation is decreasing, but the number of gypsy moth eggs is much larger than in the previous time period.
   - The infestation is increasing, but the number of gypsy moth eggs is less than in the previous time period.

Response to Calculus After High School I

Frank Wattenberg[12]

It is both delightful and difficult to comment on this paper – delightful because the paper is wonderful – difficult because finding myself in complete agreement with the substance of the paper and admiring the way in which it is written I find it difficult to add anything.

Indeed my main comment is that although the topic of Don Small's paper – a calculus course for freshmen who have completed a full year of high school calculus – is important, the reasoning and philosophy underlying the course is even more important and of much broader significance. Let's look at some of the elements of that philosophy.

Small's focus is on "student growth" and, in particular, "developing (students') ... ability to learn on their own" rather than a "felt need to expose students to the content of calculus." He then points out with extraordinary politeness the failures of the typical high school calculus course and, indeed, of the traditional calculus course still being taught on far too many campuses. "Students coming out of high school calculus courses offer a unique opportunity ... (They) usually have received a broad and procedurally oriented introduction to calculus of a single variable. Their technical knowledge exceeds both their conceptual understanding and their mathematical maturity."

I'd like to add another, more operational, description of the traditional calculus course. In the rush to cover all the topics in the typically overcrowded syllabus we wind up showing students how to solve the most routine prototypical textbook problems relying entirely on mimicry rather than thought. We do not help students develop the ability to apply a combination of mathematical ideas and techniques to complex and less-than-perfectly-posed problems in unfamiliar settings. We provide students with a tool box filled with miscellaneous tools. We produce (and the verb is significant) students skilled with saw and hammer but we do not help them become cabinet makers.

Although this characterization is probably accurate for most high school calculus courses, which after all are taught by the best high school teachers to the most successful students, it may be too generous for the typical traditional college course. I know of a recent attempt by traditionalists on one campus, to "assess" a reform course as compared to their more traditional course. The sole assessment was a one hour common, and very overcrowded, section on the final examination testing very routine skills. The underlying philosophy was that they were willing to allow their more adventurous colleagues to try Camembert provided it tasted exactly like Velveeta. The startling result of their assessment was that with one exception all the sections, both traditional and reform, did very poorly on the examination. The traditionalists had designed a test meant to show off their course and discovered that their course wasn't even achieving its goals. The one exception was a section that performed significantly better than the others and used a lot of the material and ideas from Project Calc. At this point the traditionalists rediscovered statistics and pointed out that a single experiment with small numbers and no control over self-selection might not produce worthwhile results. Nonetheless they plan to try the same thing again until they get the results they want.

Don Small also points out one of the main obstacles to change—High 'math ego' and strong loyalty to previous program. It is natural that students who have been successful as measured by high grades, have a high regard for the program in which they did well. It is also to be expected that many of these students will resist 'changing the rules of the game' .... " Of course, the problem here is more than just students. The existing faculty at most campuses have also been successful in traditional mathematics classes and have internalized the courses they took as their very definitions. When asked what calculus is or what the purpose of the course is, a surprising number of calculus instructors still respond with a long list of topics straight out of a traditional table of contents.

Small nimbly sidesteps this problem in the course he proposes – it is a different course. The content is multivariable calculus. This is an elegant solution. It provides a wonderful setting for developing the concepts of calculus and it is new. In

[12]University of Massachusetts, Amherst, MA 01003

addition, by moving multivariable mathematics forward in the curriculum we open up a much wider range of significant applications earlier.

One of the key elements of Small's discussion is the emphasis on active student involvement – "Students need to realize the primary importance of out-of-class work and the secondary importance of in-class work. We need to be clear that the purpose of class work is to support outside study and not the other way around." I would change the wording here. The important thing is what STUDENTS do, not what the INSTRUCTOR does.

Here are three tests to apply to a mathematics course.

1. If the students complain that there are not enough examples then the course is probably a good one. Students have been taught to do problems by paging backward from the exercises until they find an example to mimic. Good homework involves thought and perseverance, not just mimicry.
2. If an instructor describes what he or she is doing, then the course is probably a poor one. If the instructor brags about what the students are doing, then it is probably a good one.
3. If an instructor complains about his or her students, then the course is probably poor. If an instructor is pleased with the students, then the course is probably good. In other contexts Small has noted that his students have a lot of "street smarts." That is one of my favorite phrases. We need to capitalize on what our students do know rather than blame our failures on the students and their school teachers (who, by the way, are products of our classes).

Finally, Small emphasizes the use of real, substantial problems for two reasons. First, these problems motivate students. They stimulate an "interest in understanding as well as a need to understand. " We must engage our students in their work. Just as an English teacher would not expect to motivate grammar by studying disembodied but grammatically correct sentences like "The brindles thunked the garns bartly," we should not expect to turn our students on with endless gray lists of purely computational and meaningless exercises.

More importantly, most students study mathematics in order to be able to use it outside the classroom. Textbook problems are found only in textbooks. Outside the classroom problems are less-than-perfectly-posed. They don't come a few convenient pages after a similar example. They require several techniques not just the technique du jour. They are not expressed in exactly the right form using exactly the right words. They require modeling skills – the ability to identify what is important, the ability to express real ideas mathematically, and the ability to interpret mathematical results. Mathematical manipulation provides only one part of the solution. The best way to help students develop the ability to use mathematics to study real and significant problems is by building those problems into our courses.

Let me close with a technological note. Real problems are typically computationally complex and they require a "wet lab." For about $200, Texas Instruments' TI-92, developed jointly with the developers of DERIVE, allows students to carry a powerful CAS in their pocket. Texas Instruments is also selling their CBL, calculator-based laboratory, a marvelous tool for collecting real data about interesting questions. The last financial barriers to "real problems" have disappeared.

Response to Calculus after High School Calculus II

Jeanette Palmiter[13]

One thing that seems to distinguish experts in a field from novices, apart from the greater amount of information that the experts possess, is how that information is structured. Whereas in novices different items of information appear often to be tied together in a somewhat haphazard associate network, the information of the experts appears to be linked more in accordance with the essential organization of the subject. As the instruction seeks to add new knowledge to the memory store, it should at the same time strive to build better organized structures in memory, so that the linkages make the knowledge accessible when it is needed.

Robert Thorndike, *Intelligence as Information Processing: the Mind and the Computer, 1984*

Calculus after High School Calculus recommends a course of actions which would reduce a 3-semester calculus sequence to two semesters by concentrating on the following themes. Below I paraphrase the paper's main ideas and present my reactions:

1. The program targets the development of students' mathematical maturity by focusing on student growth rather than on content. Critical thinking and responsibility for one's own learning are needed for success in today's technological world, and the development of these characteristics needs to be central to our teaching.

   I agree. In the long run whether a topic or detail has been added or deleted from a course makes little difference. Thus we should concentrate on fostering an environment in which the student will achieve a greater degree of confidence (maturity) so they will create connections, transfer ideas, and fully understand the structure of the subject.

   In this sense, perhaps, many of the curriculum choices should be left to the individual faculty member. One teaches with more enthusiasm and with better care the topics one prefers. Thus, any proposed syllabus should serve only as a suggestion, with included topics properly justified as to their impact on the development of critical thinking in the subject area.

2. The program strives to develop a small set of concepts well so students may generalize to new situations.

   For students to attain the critical thinking and mathematical maturity stated in the first goal, the student must be able to view the subject as a series of connections between a very few basic ideas. Our traditional manner of teaching calculus has focused completely on the details and seldom the "big picture". It is no wonder students view derivatives completely from the standpoint of the algorithm of converting x^n to nx^{n-1}. We as faculty have not emphasized how the topics relate and are connected to just a handful of basic ideas.

   For example, in calculus, and especially with Taylor Series, functions are converted to polynomials. But most students do not even know what a polynomial is, nor *why* such a function is desired (although they have heard the term numerous times). By creating two lists of functions, those which are and those which are not polynomials, students can discern a pattern and develop the definition. Likewise by asking students to evaluate the two lists of functions at simple integer values, asking them to find derivatives and integrals, it soon becomes obvious as to why polynomials are preferred. The strategy of looking at examples and non-examples, deducing patterns and developing precise definitions generalizes to any new idea.

3. The program encourages students to take more responsibility for their learning and develops in students expertise to become better outside-class learners based on the premise that outside-class learning is more important than in-class learning.

   Having students take more responsibility for learning: I doubt anyone would argue against this goal. The question is how are the faculty

[13]Portland State University, Portland, OR 97207

going to turn around years of teacher-centered learning? In the paper it is suggested that certain topics will not be mentioned in class, but required in work outside of class, and then be tested. Does this approach encourage self-learning? How do the group dynamics contribute to the self-learning?

The paper fails to mention if the model class is taught in a lecture mode or a variation. It is acknowledged the class meets about once a week in the computer lab. I have also found group activities in the lab to be successful when students are placed in pairs. Larger group sizes are not conducive to full participation.

I disagree with the premise that in-class learning is less important than out-of-class learning. Class time is very valuable for putting the pieces together, interpreting the situation, and mapping out where one is headed. The perspective is created and discussed. Out-of-class time works through the details.

I do not allow my students to take notes in class, or only a very minimal amount (not more than one page). At first this makes many of them nervous. They are accustomed to frantically copying notes off the boards or scribbling teacher's comments, yet by doing so, the student is not able to fully participate in the discussion, nor able to follow, extend, or debate arguments. Active learning in class replaces passive note taking and the notion that one can learn better by "second-hand" notes at a later time. The textbook serves as the set of notes.

The no-notes approach has worked very successfully in my classes, even those with large lecture numbers. By constantly questioning my students I discover their level of understanding and their misconceptions, keep them informed about the big picture and how it relates to what we have done and what we will be doing. I keep them perpetually engaged and they are not hesitant in asking questions or providing many insightful "what if" queries. Out-of-class time is spent sorting out the details and considering special cases. Working in groups is extremely useful for out-of-class assignments.

4. The paper suggests 10 topics to delete or de-emphasize which include graphing rational functions, Newton's Method, volumes and surfaces by revolution, minimal coverage on techniques of integration, polar/cylindrical/spherical coordinates, minimal coverage on convergence tests for series, spreading sequences over the course and omitting lone treatment, hyperbolic functions, conic sections, Stokes and Divergence theorems.

I agree with deleting or de-emphasizing most of the ten topics. With the available technology the algorithms and algebraic tricks we drilled our students on in the past are not valued any more.

Many of the calculus reform papers list deletions. More important is what is to be kept. The paper lists a syllabus, but justification is lacking. Why is a certain topic retained or why is another enhanced? The paper recommends l'Hopital's Rule be retained, even though with the advent of computer algebra systems this rule is just another "trick" which perhaps could be omitted. I personally still teach it because it is such a clever idea, and I just enjoy teaching it. But the reasons for keeping a topic should correlate with how the topic fits in with the subject or lends itself to clearer understanding of basic concepts.

As was mentioned earlier in the review, we should be striving to provide a conceptual and applicable calculus course. Deleting or adding a topic such as l'Hopital's Rule will not drastically affect a student's understanding of calculus. Many of the choices should be made by the individual faculty member. Many faculty are looking for ideas based on justified recommendations that are convincing. Our suggestions on which topics to keep may need to begin with why should we teach differentiation and integration.

5. The paper recommends a generalized and two-semester approach to those students who have *successfully* completed high school calculus.

The suggestions and ideas espoused in the paper would also serve audiences that were not "successful" in high school calculus or those who are taking calculus for the first time. For those students who were not initially successful in high school calculus, this presentation may prove more meaningful. For those learning it for the first time, why not present it correctly at the start? Let's not restrict these good ideas to only the successful high school calculus students. Likewise, the ideas should not be limited to the college level; high schools may wish to adopt the approach, too.

6. By presenting calculus through an integrated approach that emphasizes generalization, students will be led toward conceptual understanding and involvement in discovery and exploratory work, which in turn accelerates the growth of mathematical maturity.

   The expectation is that the proposed changes will lead to payoffs. It would be useful, and more convincing, if the paper had cited evidence of how the integrated, core approach has been beneficial, even if the evidence was anecdotal. If an improved difference has been noted, are the results due to the new syllabus, the textbook, the technology? Or are the favorable results mainly due to the efforts of the faculty member? If so, this in no way diminishes the outcomes nor does it mean the program cannot be duplicated, but instead it supports the notion of allowing faculty freedom in the classroom.

The greatest impact of the calculus reform movement and the use of technology in the calculus classroom has NOT been on the students but on the faculty (with the students reaping many of the rewards). Due to the reform, faculty are now questioning what has been traditionally taught, examining their teaching strategies, and striving to improve the experience for the students. The movement has transformed "burnt-out" professors into ones who truly enjoy teaching. The "7-into-4" curriculum plans, if they do nothing else, will keep faculty focused on the way we teach our crucial beginning courses.

Linear Algebra in the Core Curriculum

Donald R. LaTorre[14]

How can we incorporate the essential content of the sacred seven - Calculus I & II (single variable), Calculus III (multivariable), Differential Equations, Discrete Mathematics, Linear Algebra, and Probability/Statistics - into four courses and still provide a unified and coherent curriculum? How can we design a curriculum that is interesting, relevant, and which does not short-change the long-term needs of students from an increasingly diverse group of client disciplines?

Linear Algebra in Undergraduate Mathematics

The importance and role of linear algebra in the undergraduate curriculum has been articulated carefully by Alan Tucker [11]. Designers of any core curriculum are well-advised to spend some time assimilating Tucker's elegant article.

In a world that mathematicians tend to regard as consisting of a vast number of complex systems with many interacting variables, linear models have become a primary and important tool. Linear Algebra, with its rich tradition of matrix representations, notational elegance, and vector space outlook, coupled with its more recent development along algorithmic, numerical, and computational lines concurrent with the growth of computing over the last 30 years, is unique in its potential to clarify, simplify, and unify a large number of the linear models that we teach to undergraduates. The models range from those employing the elementary vector methods of multivariable calculus to more sophisticated ones represented by systems of linear differential equations, linear difference equations and dynamical systems, input/output systems in economics, Markov chains, constrained optimization, graphs and networks, and linear statistical models. Since any core curriculum program will almost certainly incorporate elements of linear modeling, it is essential that a firm grounding in the fundamentals of linear algebra - both theoretical and applied - be included. From the point of view of linear algebra, we must give up the traditional sophomore course in the subject and zero in on the content, context, spirit, and methodology that is most appropriate for a core curriculum.

Context

Because of the growing importance of modern matrix methods in many applications of mathematics, and the easy access to computation with improved matrix algorithms provided by the products of modern technology, it is clear that the linear algebra in any core curriculum should be developed in the context of matrices. The traditional emphasis on vector spaces, linear transformations, and other such abstractions should give way to a more practical, problem-solving approach with plenty of motivational examples.

This is precisely the approach that has been recommend by the Linear Algebra Curriculum Study Group (LACSG), a broadly-based NSF supported project to initiate substantial and sustained national interest in improving the undergraduate linear algebra curriculum. Their recommendations regarding the first course in linear algebra can be found in the article by David Carlson et al.[1], and state, in part: "Mathematics departments should seriously consider making their first course in linear algebra a matrix-oriented course. (The) course should proceed from concrete, and in many cases practical, examples ...."

There are ample teaching materials that reflect this point of view. Perhaps the best known (but not new) is Gil Strang's classic text [9], which adopts a matrix-theoretic approach and chooses "*... to explain rather than to deduce*". Strang's new text [10] carries this approach one step farther at an even lower level. Alan Tucker's fine texts [12], [13] are especially unique in that they introduce students to the main concepts in a matrix-theoretic treatment through the repeated use of a few linear models. This may be an optimal way to approach linear algebra in a core curriculum. David Lay's text [7] is probably the first one to embrace the LACSG's philosophy to the fullest extent. It includes a wide range of applications that illustrate the power of linear algebra in a diversity of fields. Charles Cullen's text [2] is noteworthy because it presents a fully integrated treatment of linear algebraic systems and linear differential equations.

[14]Clemson University, Clemson, SC 29634

Content

In terms of content, there is a large list of topics in linear algebra from which to choose. We must decide what is central to our purposes and what is not, what we really want students to know and to be able to do. In short, we must decide what we *ultimately value* in their learning of linear algebra and let go of the rest.

Although there are no easy answers to the content issue, there is at least a good starting point. Among the recommendations of the LACSG is a core syllabus that can be covered in 26-28 fifty-minute classes. Since there are usually 57-58 such classes in a typical 15-week semester in which classes meet four times each week, this core syllabus requires roughly one-half a semester to complete, likely all the time that can be devoted to linear algebra in a core curriculum. If classes were to meet five times each week, an additional 7 fifty-minute classes would be available. The core syllabus represents the best thinking of a dedicated group of seasoned professionals with considerable input from consultants in client disciplines, and was refined over a 2-year period.

The LACSG Core Syllabus

I. MATRIX ADDITION AND MULTIPLICATION (3 DAYS)

This includes the normal topics of matrix addition, scalar multiplication, matrix multiplication, transposition, and their algebraic properties such as associativity of matrix multiplication. Operations with partitioned matrices. Motivate matrix multiplication and carefully examine three views of the product AB:

1. Ax is a linear combination of the columns of A, with coefficients from x; each column of AB is obtained by multiplying A by the corresponding column of B. Thus, each column of AB is a linear combination of the columns of A, with coefficients from the corresponding column of B. If D is a diagonal matrix, then AD is a scaling of the columns of A. If P is a permutation matrix, then AP is a permutation of the columns of A.
2. Similarly, the rows of AB are linear combinations of the rows of B.
3. AB is a sum of outer products (i.e., rank 1 matrices): $AB = col_k(A)row_k(B) + \ldots + col_1(A)row_1(B)$, when A is m by k and B is k by n.

II. SYSTEMS OF LINEAR EQUATIONS (4 DAYS)

Gaussian elimination/elementary matrices. Echelon and reduced echelon form. Existence/uniqueness of solutions. Matrix inverses. Row reduction interpreted as an LU-factorization.

III. DETERMINANTS (2–3 DAYS)

Determinants are readily encountered when solving 2 by 2 and 3 by 3 general linear systems. The elementary properties of determinants are easily discovered or illustrated using the resulting expressions. Formal verifications in most cases should be avoided. Explore the uses of determinants as well as the difficulties in computing them. Main topics: cofactor expansion, determinants and row operations, $detAB = detAdetB$, and Cramer's Rule (to show the sensitivity of solutions to $Ax = b$).

IV. PROPERTIES OF R^n (7–8 DAYS)

Introduce R^n as a set of n-tuples and not as a formal vector space. Define vector addition and scalar multiplication, but it is not necessary to prove formally all the properties of vector addition and scalar multiplication. There should be a strong geometric emphasis in the presentation of this material.

1. Linear combinations: linear dependence and independence.
2. Bases of R^n.
3. Subspaces of R^n: spanning set, basis, dimension, row space and column space (range of A as a mapping), null space.
4. Matrices as linear transformations.
5. Rank: row rank = column rank, products, connections with invertible submatrices.
6. Systems of equations revisited: solution theory, rank + nullity = number of columns.
7. Inner product: length and orthogonality, orthogonal/orthonormal sets and bases, orthogonal matrices.

V. EIGENVALUES AND EIGENVECTORS (6 DAYS)

Eigenvalues are important in a wide variety of applications. Sufficient time should be allowed for complete coverage of this topic. Eigenvectors may be introduced and/or motivated using geometric examples.

1. The equation $Ax = \lambda x$.

2. The characteristic polynomial and identification of some of its coefficients (e.g. trace, determinant), algebraic multiplicity of eigenvalues.
3. Eigenspaces, geometric multiplicity.
4. Similarity: distinct eigenvalues and diagonalization (with emphasis on $AP = PD$).
5. Symmetric matrices: orthogonal diagonalization, quadratic forms.

VI. More on Orthogonality (4 days)

Include the standard topics with a strong geometric emphasis: orthogonal projection onto a subspace; Gram-Schmidt orthogonalization and interpretation as a QR factorization; and the least square solutions of inconsistent linear systems, with applications to data-fitting.

Total: 26–28 days

Comments on the Core Syllabus

In considering this core syllabus, there are several points to keep in mind.

1. The syllabus was designed to be the core of a semester's course in introductory linear algebra at the sophomore level with the prerequisite of a full year's study of calculus. To implement it in any reasonable way into a core curriculum, where the stated prerequisite may no longer hold, will require considerable care. Since there are likely to be several versions of any such core curriculum, each with its own unique emphasis, the core syllabus should also be examined to determine the extent to which each of the suggested topics supports the desired overall emphasis.
2. Intended as part of a more extensive course in linear algebra, the core syllabus reflects the main themes in such a course: matrix operations, systems of linear equations, properties of R^n, etc. But in any core curriculum, it is likely to make more sense to introduce the individual topics from the main themes at varying points in the curriculum, with each new topic supporting one or more specific applications, and often foreshadowing future material.

   As an example, we have considerable experience from the teaching of multivariable calculus that suggests that we should not introduce R^n in its full generality at the outset, but instead begin with elementary vector concepts in R^2 and apply them to vector-valued functions and motion in the plane, then revisit the same concepts in R^3 with applications to motion in space. The more general development of R^n can wait until later. But this approach is likely to add another 2 days to the syllabus.

   We can foreshadow the development of R^n as a vector space by being careful to note the eight defining properties (four for vector addition and four for scalar multiplication) initially for vectors in R^2, then for vectors in R^3, and again later for addition and scalar multiplication of matrices. And we can foreshadow the troublesome concepts of linear dependence and independence of vectors in R^n early in a study of linear systems by interpreting each non-zero solution to a homogeneous linear system $Ax = 0$ as a linear dependence relation among the column vectors of matrix A. Linear independence can likewise be first introduced in terms of the trivial solution to $Ax = 0$.
3. The core syllabus does not explicitly mention cross products, scalar triple products, and lines and planes in R^3- all topics with important applications in multivariable calculus. To include these topics will require another 2 days. We have now added a total of four days, so consider reducing the treatment of determinants to a bare 2 days. We now have 30-31 days, so five class meetings each week seems to be optimal.
4. Abstract vector spaces are deliberately avoided, but it would seem imperative to at least mention the following as important examples of spaces of vectors: the space R^∞ of all infinite sequences, the space $R^{m\times n}$ of all real $m \times n$ matrices, the space $F[a, b]$ of all real-valued functions defined on the interval $a \leq x \leq b$, and the space $R[x]$ of all real polynomials of finite degree. Likewise, the derivative and integral operators should be recognized as examples of linear transformations on elementary function spaces.

The Role of Technology

Just as computing is helping to reshape and redefine the practice of linear algebra at the professional level, so also is it affecting major change in the teaching and learning of linear algebra at the undergraduate level. Computing, either with microcomputers or high-level graphics calculators, should therefore be an *essential ingredient* of linear algebra in a core curriculum, and has been strongly recommended in both [1] and [3]. Its careful use can inject a new spirit of inquiry and discovery, ease the computational burden associated with hand execution of matrix algorithms, facilitate more realistic applications, and increase awareness of some of the real computational issues. Computing should be an integral part of the development, not an isolated component, but programming should be avoided. The new fourth edition of Steven Leon's text [8] is representative of the new breed of linear algebra textbooks in this regard; it contains a wealth of MATLAB computing exercises at the end of each chapter.

The real question is how much computing, and when. The severe compression of material, even with five classes per week, is likely to leave little time for formal laboratory work. Thus, high-level graphics calculators such as the HP-48G/GX and the TI-85, which are being widely embraced throughout the first two years of undergraduate mathematics, may offer the most reasonable promise for computing in a core curriculum. They are now fairly sophisticated in their matrix routines, and often rival many microcomputer systems. Teaching code for the HP-48G/GX series calculators can be found in [6] and for the TI-85 units in [5].

References

[1] Carlson, David, et al, "The Linear Algebra Curriculum Study Group Recommendations for the First Course in Linear Algebra", *The College Mathematics Journal*, 24 (1993), 41-46.

[2] Cullen, Charles G., *Linear Algebra and Differential Equations*, Second Edition, PWS-Kent Publishing Co., 1991.

[3] Herman, Eugene, et al, "The Use of Computing in the Teaching of Linear Algebra", *Computers and Mathematics*, MAA Notes No. 9, The Mathematical Association of America, 1988.

[4] LaTorre, Don, "Using Graphing Calculators to Enhance the Teaching and Learning of Linear Algebra", *Symbolic Computation in Undergraduate Mathematics Education*, MAA Notes No. 24, The Mathematical Association of America, 1992.

[5] LaTorre, Don, "Explorations in Linear Algebra", Chapter 6 in *Explorations with the Texas Instruments TI-85*, (John G. Harvey and John W. Kenelly, Editors), Academic Press, 1993.

[6] LaTorre, Donald R., *Linear Algebra: Teaching Code for the HP-48G/GX*, Saunders College Publishing, 1994.

[7] Lay, David C., *Linear Algebra and its Applications*, Addison-Wesley Publishing Company, 1994.

[8] Leon, Steven J., *Linear Algebra with Applications*, Fourth Edition, Macmillan College Publishing Company, 1994.

[9] Strang, Gilbert, *Linear Algebra and its Applications*, Third Edition, Harcourt Brace Jovanovich, 1988.

[10] Strang, Gilbert, *Introduction to Linear Algebra*, Wesley- Cambridge Press, 1993.

[11] Tucker, Alan, "The Growing Importance of Linear Algebra in Undergraduate Mathematics", *The College Mathematics Journal*, 24(1993), 3-9.

[12] Tucker, Alan, *A Unified Introduction to Linear Algebra: Models, Methods and Theory*, Macmillan Publishing Company, 1988.

[13] Tucker, Alan, *Linear Algebra: An Introduction to the Theory and Use of Vectors and Matrices*, Macmillan Publishing Company, 1993.

Response to Linear Algebra in the Core I

David Carlson[15]

I think that Don LaTorre has put the case for linear algebra quite well. As they say in courtroom movies, I think that the case rests, with his presentation of:

- the pervasiveness of linear algebra throughout mathematics and its applications;
- the value of a concrete approach, which I note can start in $\mathbf{R}^2$and $\mathbf{R}^3$, demonstrate the close relationship between linear algebra and geometry, and then show how matrices and linear algebra can simplify and clarify ("roll back the fog", if you will) and geometrically represent problems in any number of variables; and
- the value of the syllabus proposed by the Linear Algebra Curriculum Study Group (LACSG), as a carefully-designed first approximation to an answer to the question, "What linear algebra content do we want our students to master in their first four semesters?"

The development of the LACSG recommendations with their proposed syllabus for elementary linear algebra are a much simpler version of what we are attempting here. The LACSG recommendations try to identify what is important in content and approach; the syllabus proceeds from what is concrete for our students (algorithms like matrix multiplication and Gaussian elimination) to what is abstract for them (concepts like spanning and linear independence), all in the contexts of $\mathbf{R}^n$and matrices. (The axioms for a vector space are optional.) The LACSG recommendations also recognize the need for further study into pedagogical issues like student learning (and its assessment) and instructional innovation (and its evaluation).

I would like to comment on the very useful second question proposed by the organizers of this meeting, that of the integrative threads which run through the "Sacred Seven," and of their role in possibly enabling the replacing the seven by a "Final Four". I would like to mention some threads of content.

One such thread that I see is geometry, which of course is not one of the Sacred Seven. I tell my students in my linear algebra classes that often they can see what to do with geometry, and then use linear algebra to actually do it. This is certainly true also in calculus. I think that one might well make a third course in calculus into a course on geometry and linear algebra and calculus in $\mathbf{R}^2$and $\mathbf{R}^3$. Then one can bring in linear geometric maps, projections, reflections, and rotations, and discuss their real eigenvalues and eigenvectors. This allows one to see how the eigenvectors help us think about the maps in useful ways. It would also be possible to discuss orthonormal bases, and how they help us. Here I'd like to mention the book by Tom Banchoff and John Wermer, *Linear Algebra Through Geometry*, which does $\mathbf{R}^2$and $\mathbf{R}^3$ linear algebra very completely before going on to $\mathbf{R}^n$.

Another thread I see is the use of matrices in linear geometric maps and in systems of linear equations and of linear differential equations as well as in incidence matrices and elsewhere. Having put two– and three–dimensional linear algebra into that third course of calculus, one could usefully put much of the rest in with differential equations, as done in the Cullen book Don LaTorre mentioned.

I'd like to mention one more thread, the role of indices and their use in summation and induction. This thread appears most clearly in discrete mathematics and in matrices. Perhaps systems of linear equations and matrix operations could fit together with discrete mathematics. Indices seem so easy to us, but working with unspecified indices is not easy for our students.

Let me conclude with two ideas about our work here in general. The first is that our Core Curriculum will not be complete without an appropriate common computing environment. Whether it involves a supercalculator or a software package like MATLAB or Maple, we should decide what will work for all the ideas and techniques of the first two years.

I also like the idea of thinking of our goal as a Core Curriculum, rather than as 7-into-4. The concept of 7-into-4 sounds as if we were trying to stuff the contents of seven suitcases into four. We might be able to succeed with suitcases, but the results in terms of student learning are likely to be

[15]San Diego State University, San Diego, CA 92182

disastrous for mathematics courses. If our project is to succeed, we must end with four courses which are conceptually and pedagogically sensible individually and as a package, and which contain no more material than our students can assimilate in four semesters. The writers of the syllabus and course materials for these four semester courses will have to try them out successfully with a variety of student populations before we will know that we have reached our goal.

Response to Linear Algebra in the Core II

Anita E. Solow[16]

In thinking about the role of linear algebra in the first two years, one often starts by listing the topics that must be taught. The LaTorre paper contains such a syllabus. Unfortunately, this process leads to a long list of mandatory topics, without reasons given to why those topics should be taught at all, why they belong in the first two years, and how they relate to other mathematics that should also be in the first two years. We know why these topics have been taught in the past, but perhaps it is now time to rethink the courses from the beginning, starting with declaring nothing sacred (even our favorite topics).

I teach at Grinnell College, a private liberal arts college in Iowa. We have pressures on us to teach a large amount of mathematics in few courses. Our students need only take 8 courses for a mathematics major, and taking more than one mathematics course at a time during the first two years is not usually encouraged. Therefore, we have thought quite a bit about what should be taught in mathematics for the first two years. I should mention that in one important way the situation at Grinnell is very different from the one here at West Point. We have no required courses. So in our planning, we cannot assume that students in quantitative majors other than mathematics will take four mathematics courses.

We have ended up with what, in the spirit of this conference, one could call a "5-into-3" solution. We teach two semesters of calculus. The first semester is single variable and includes derivatives and integrals. The second semester focuses on multivariable calculus. In order to do this, the study of series is postponed until a later course. In the first semester of the sophomore year, students take a combined linear algebra/linear differential equations course. The calculus year works well and is considered a successful sequence of courses. It is the linear algebra course on which I want to concentrate here.

We have been teaching this course for about a dozen years. The bottom line on this course is that it does not work. The student grapevine has declared that this is a "killer" course, and perfectly able students fear taking it. We have tinkered with the course, but there is still a sense of dissatisfaction with it. Why? The philosophy of the course makes wonderful sense for a mathematician: the students study linear algebra, including abstract vector spaces, and then immediately apply this theory to the solution of linear differential equations. The major problem with this philosophy is that the students have not grasped the theory well enough to apply it to anything. It is true that some of the better students are helped in their understanding by the application to differential equations, but for the majority of students, it all remains a mystery.

So, I offer my first warning. Students need time and experience to make sense of hard topics. Integration of topics that may make sense to us as mathematicians may not be successful in the classroom.

There are two other reasons why we are not satisfied with the linear algebra/linear differential equations course that are pertinent to this conference. One has to do with calculus reform. We learned in changing calculus that we needed to throw out topics that were not central to the course. In the place of the omitted topics, we did not add new topics. Rather we left time for labs, projects, group work, etc. All of these activities take time. We would like to continue the pedagogy that we have developed for calculus into the next course, but this puts pressure on us to pare down an already pared down course.

I cannot imagine teaching linear algebra without technology. I remember how horrible it was to grade students' work when it was all done by hand. There was a great deal of time and effort on the part of the students on computation, often obscuring the ideas of the course. But although technology has provided time in the course, it has also taken it away. To be sure, since we use the computer in linear algebra, students no longer spend time doing matrix operations by hand, and we no longer spend class time discussing the fine points of Gaussian Elimination. However, technology allows students to explore relationships and make conjectures and to examine realistic applications of linear algebra. Students can deal with Leslie population models and use real data. They can analyze the economy of a

[16]Grinnell College, Grinnell, IA 50112

country using the Leontief model. But these take more time than is saved by eliminating the computation.

So, my second warning is that changing pedagogy and technology both put pressures on a course to increase the amount of time needed for studying a topic. These pressures run counter to the effort to teach more mathematics in fewer semesters.

I do not wish to be overly pessimistic, only realistic. The 7-into-4 problem will require us to rethink the entire first two years, and not from the perspective of the individual courses. Instead we need to balance the demand for new topics, technology, and active pedagogy against the traditional curriculum that is in place largely for its historic use in the physical sciences.

Response to Linear Algebra in the Core III

Steve Friedberg[17]

While examining Don LaTorre's recommendations regarding linear algebra in the 7-into-4 Curriculum and how they would affect mathematics majors as well as those from the client disciplines, I focused on the following considerations:

- The connection between linear algebra and the rest of the mathematics curriculum – How should linear algebra prepare mathematics majors for the abstraction in later courses? How much rigor is expected?
- Topics – What's in and what's out?
- Chronology –Is the order of the topics reasonable?
- Time allotments – Is there a sufficient amount of time assigned to each topic?
- Writing assignments – How much and what level of writing should we demand?

I believe that linear algebra should serve as a bridge to later courses within the mathematics major. I agree with Alan Tucker (1993) who notes that "... since calculus is taught as a service course with little theory, a balancing theory orientation to linear algebra appears vital for mathematics majors advancing to courses in abstract algebra and analysis." Likewise, the Linear Algebra Curriculum Study Group (1993) notes that the proposed course should contain no "less rigor or theorem proving" than the traditional one. It is with this expectation in mind that I examined the chronology and time allotments.

I do have a number of points of agreement with the general philosophy and the choice of topics expressed in LaTorre's paper. For example, I agree that a much more matrix oriented approach is needed with perhaps an almost exclusive emphasis on $\mathbf{R}^n$. Block operations provide several views of matrix multiplication and should be used to see the relationship between linear independence, the rank of a matrix, and the existence of nontrivial solutions to a homogeneous system of linear equations. Of course, numerical methods and technology (through either hand calculators or personal computers) should be integrated wherever they make sense. Whenever possible, geometric applications and interpretations should be given. Finally, a diverse collection of applications to the real world as well as mathematics should be interspersed throughout. I agree that abstract vector spaces should be avoided within a first course in linear algebra. However, as the paper notes, in a 7-into-4 curriculum it will be necessary to introduce the important function spaces of analysis.

Some appropriate topics that should be added are: elementary iterative methods for solving systems of linear equations (e.g., Jacobi and Gauss-Seidel methods), eigenvalue location (Gerschgorin's disk theorem) and approximation, and systems of difference and differential equations.

What follows is a list of suggestions that I believe may be improvements or enhancements to the recommendations.

1. The hard topics should be introduced earlier and should be given more time. As proposed, the beginning of the course (Sections I, II, and III) seems too heavy on computation (matrix arithmetic, Gaussian elimination, and cofactor expansion). As a result, Section IV follows with linear independence, subspace, spanning set, basis, and dimension in rapid succession. These topics along with rank, systems revisited, and orthonormal bases are to be done in seven or eight days. I believe that a reasonable understanding of these topics in this compressed time is beyond most sophomores, particularly considering the previous LACSG statement concerning rigor. In my experience not only at Illinois State University, where we allot over twice as much time for this material, but even at MIT, students have great difficulty even describing, let alone proving, the simplest results about spanning sets and linear independence. With the 7-into-4 Curriculum, the students will apparently have even less prerequisite college mathematics than we assume now.

   I recommend not only more time be spent on these topics, but that at least some of the topics appear much earlier. For example, subspaces such as lines and planes through the

[17] Illinois State University, Normal, IL 61790

origin as well as their spanning sets can be introduced any time after vector operations in $\mathbf{R}^n$ are defined. The spanning set of the solution space to a homogeneous system obtained by Gaussian elimination, as well as the column space of a matrix can be discussed immediately after matrix operations are defined. As Don LaTorre notes, linear independence can be discussed in terms of a nontrivial solution to a system of (linear) equations; so, these topics can appear in Section II.

2. Linear Transformations from $\mathbf{R}^n$ to $\mathbf{R}^m$ should be interwoven throughout.

   The recommendations relegate linear transformations to the supplementary topics. Unlike abstract vector spaces, functions have been the mainstay of the previous (calculus) courses and so are much more familiar to students. Linear transformations are not generalizations of matrices in the same sense that abstract vector spaces are generalizations of $\mathbf{R}^n$; indeed, they are intimately connected with matrices. The use of phrases such as "range of a matrix" and "matrices as linear transformations" within the recommendations indicates the need for introducing linear transformations per se. The application of such familiar properties as one-to-one and onto within the context of the uniqueness and existence of solutions to systems should be natural to the student. The quote by P.R. Halmos, "Matrix multiplication is an algorithm – but do you remember the proof that it is associative?" comes to mind as I argue for a stronger emphasis on linear transformations. LaTorre states the desirability of noting that the derivative and integral are examples of linear transformations. This makes more sense if the student is already familiar with linear transformations.

   I list several additional reasons for including linear transformations. (The notation L_A denotes the left-multiplication transformation, that is, $L_A(x) = Ax$, where A is matrix and x is a vector.)

   - It makes intuitive sense to study geometrical concepts such as rotation, reflection, projection, and isometry as linear transformations. Likewise, eigenvalues and eigenvectors are easier to motivate from the geometrical viewpoint of stretching or compressing vectors rather than as solutions to the equation $\det(A - \lambda I) = 0$ or the system $(A - \lambda I)x = 0$.
   - Many statements may be viewed either in the language of matrices or linear transformations. The students should have access to both. Compare the equivalent statements that guarantee the system $Ax = b$ has a solution: (i) b is a linear combination of the columns of A and (ii) b is in the range of L_A.
   - Linear transformations are coordinate-free unlike multiplication by a matrix, and so they are easier to generalize.
   - Obtaining an algebraic formula for a geometric transformation is made easier when the transformation is thought of as a linear transformation rather than as multiplication by a matrix. For example, suppose that the student wants to find the algebraic formula that represents reflection of R^2 about the line $y = 3x$. With linear transformations available, the student immediately identifies the basis $(1, 3), (-3, 1)$ as an appropriate set of eigenvectors of the associated linear transformation. The matrix representation is a diagonal matrix, from which the standard matrix and hence the algebraic formula is quickly computed.

3. Writing should play a significant role.

   In every major mathematics course at Illinois State University, students are given Guidelines for Writing Mathematics – a list of rules for good mathematical writing. In the sophomore linear algebra course students are given weekly written assignments. For example, they may be asked to prove or disprove that a given subset of $\mathbf{R}^n$ is a subspace, that a subset of a linearly independent set is linearly independent, or that the trace of AB equals the trace of BA. This work is carefully graded and has a significant impact on the course grade. It is expected that the exposition be of sufficient clarity that other students in the class could profit from the explanations. Careful writing at this stage makes the transition to the upper division courses easier.

In summary, I very much like the increased emphasis on matrices and technology. Linear transformations play a very important role in the first course and should be introduced early and used throughout. The harder topics require more time and less clustering. Finally, required writing assignments not only increase the students' understanding of linear algebra, but also are important as an end in themselves.

References

[1] Carlson, David et al, "The Linear Algebra Curriculum Study Group Recommendations for the First Course in Linear Algebra," *The College Mathematics Journal*, 24 (1993), 41-46.
[2] Steen, Lynn A. (Ed.), "Reshaping College Mathematics," MAA Notes, Number 13, 1989.
[3] Tucker, Alan, "The Growing Importance of Linear Algebra in Undergraduate Mathematics," *The College Mathematics Journal*, 24 (1993), 3-9.

Differential Equations in the Core Curriculum

James V. Herod[18]

Introduction

It would be presumptuous for me to suggest that all other topics in the first two years of undergraduate mathematics are pointed toward giving an understanding of the notions associated with a study of differential equations; that all else we talk about is a part of the preparation for understanding the important ideas in differential equations. I acknowledge that this would be a biased perspective.

It has been an interesting idea to incorporate seven parts of the undergraduate curriculum into two years. The calculus, linear algebra, differential equations, discrete mathematics, probability, and statistics are woven together into the program called "7-into-4." Previously, the model for a traditional sophomore differential equations course was different. The course stood at the intersection of multiple paths: a path from the calculus and paths into and out of mathematics from the sciences and the engineering disciplines. These paths merged and moved toward applications of mathematics, or toward mathematical analysis. One could view much in analysis as an attempt to make precise what had been suggested in a study of differential equations, or as preparation for better tools–or other tools–to solve partial differential equations and other dynamical systems.

This view of differential equations as the center of the first two years of mathematics is the perspective of the undergraduate curriculum for a science or engineering student and for an institute of technology. Maybe this should also be the view for a student of public policy, a student of biology, or a student of economics as we plan a curriculum. Every area of modern sciences relies on quantitative concepts to analyze phenomena, explain mechanisms, and describe the state of knowledge of complex systems.

For example, recently, there was a seminal paper titled *Mathematical Models and the Design of Public Health Policy: HIV and Antiviral Therapy*[2]. This paper addressed important issues that decision makers in public health policy must consider. The foundations of that paper form a beautiful example of the heart of our discussions at this conference. It involves the gathering and understanding of data, the construction of a model, and an analysis of the resulting model. A variety of tools were used in the steps toward developing that analysis. Finally, the model was formulated as a system of differential equations.

As they talk to the mathematics faculty, our colleagues in the program in public policy emphasize that they want their students to have strong quantitative skills and quantitative understanding. As another example of the awareness of the need for quantitative studies in diverse areas, we now have at Georgia Tech an undergraduate course in mathematical biology that runs each year and is populated with not only mathematics and biology students, but also students from earth and atmospheric sciences, health physics, and from across the campus.

The question we discuss here is certainly not about the importance of the differential equations, but more about how it should be interwoven with the other six components of "7-into-4." Most differential equations that arise in science and engineering are nonlinear, and often do not lend themselves to solutions in terms of the elementary functions. Moreover, these equations are not one-line equations, but are nonlinear interminglings of derivatives of several functions. A common theme for understanding a nonlinear system is to approximate the system with an easier linear system.

Thus, a first goal in a study of undergraduate differential equations is to make a considerable effort toward understanding a framework for linear problems and toward understanding their solutions. Then one must consider nonlinear problems. The analysis in such a collection of ideas at this level could come in these steps: approximate the nonlinear problems with linear ones, worry about the accuracy of the approximation, and hope to predict the behavior of the nonlinear problems by understanding the behavior of these linear approximations.

I argue that there are several goals for differential equations in the first two years of university mathematics.

[18]Georgia Institute of Technology, Atlanta, GA 30332

1. Linear differential equations should be studied in the context of linear algebra.
2. The concept and computation of the exponential of a matrix should be a central method used in understanding solutions of linear problems.
3. Connections should be made between the interplay of the location of the eigenvalues and the invariant subspaces of the matrix A with the solutions of the differential system $Z' = AZ$.
4. Nonlinear systems should be approached by making linear approximations.
5. The long range forecast and stability for solutions should be viewed as likely more important than the transient solutions.
6. Discrete and continuous models should be compared and contrasted.
7. Take note that some classical equations arise so often that they are a part of the heritage of sophomore differential equations.

Finally, there are several points that I want to make about the use of the technology in undergraduate education, and especially in undergraduate differential equations:

1. Technology allows solutions to be computed that were nearly inaccessible to undergraduates previously.
2. Technology allows us to compute and graph numerical solutions on a single platform.
3. Technology allows us to see patterns, instead of just the details.

The Important Ideas for Differential Equations in the First Two Years of Mathematics

In making this kneading of differential equations into the mix of ideas that is called "7-into-4" at the United States Military Academy, one should think seriously about what the students in a traditional sophomore differential equations course are expected to know. Then one should ask two questions:

1. Are those expectations still appropriate midway through the last decade of the twentieth century?
2. Are the important ideas in mathematics that arise in differential equations a part of the "7-into-4" program?

IMPORTANT IDEA I: LINEAR DIFFERENTIAL EQUATIONS SHOULD BE STUDIED IN THE CONTEXT OF LINEAR ALGEBRA

All across the country, students are being drilled on how to solve second order, constant coefficient differential equations, with a forcing function that is some combination of sines, cosines, and exponentials. The germ of an idea here involves the character of the roots of the associated polynomial. This problem is given so much emphasis for three reasons. First, it is a special case of nth order equations with constant coefficients. Second, humans can factor the associated polynomial. And third, there is a natural association with second derivatives and studying the rate of change of rates of changes in physical phenomena.

The important idea to mark, however, is that while second order equations can be understood in terms of the elementary functions, the appropriate context for the equations is their imbedding as a two dimensional system. For example, these problems are equivalent:

$$y'' + ay' + by = 0$$

and

$$\begin{pmatrix} u \\ v \end{pmatrix}' = \begin{pmatrix} 0 & 1 \\ -b & -a \end{pmatrix} \begin{pmatrix} u \\ v \end{pmatrix}$$

The eigenvalues of this matrix are precisely the roots of the quadratic equation associated with the second order equation.

It is also good to see that, just as some systems seem to be formulated most naturally as a second order equation and can be re-written as a two dimensional system, so too there are problems that arise as two dimensional systems that can be changed to second order equations.

The following equations form a linearization of what happens in a glucose tolerance test. The first equation models how the amount of glucose in the blood stream decreases due to tissue uptake and storage in the liver and how the amount of glucose decreases due to the presence of the hormone insulin. The second equation models how the amount of insulin increases due to a surplus of glucose and decreases with a surplus of insulin.

$$\begin{aligned} \frac{dg}{dt} &= -ag - bh \\ \frac{dh}{dt} &= cg - dh \end{aligned}$$

The constants are determined by drawing blood at some time intervals after the subject has drunk a notoriously sweet concoction. It is an interesting exercise to ask how often and how many blood samples must be drawn to determine with sufficient accuracy these constants. But the point here is that this system can be changed to a familiar second order equation:

$$\frac{d^2g}{dt^2} + 2a\frac{dg}{dt} + \omega^2 g = 0$$

This is the pattern. What worked here for a first order, two dimension system connecting with a second order equation carries over for first order, n dimension systems and n^{th}order equations. The associated polynomials are the same. The roots of this polynomial generate the solution and the position of the roots in the complex plane determines the character of the solutions.

Linear algebra must be an accessible tool for differential equations. This accessibility should include an understanding about the importance of eigenvalues for a matrix. A part of that understanding is the computation of the Jordan Canonical Form for a matrix. In that representation of a matrix A, one finds projection matrices P_i, and nilpotent matrices N_i so that the matrix A can be represented as a linear combination of these. For example, perhaps one can establish that

$$A = 2P_1(I + N_1) + 3P_2$$

Here, P_1 and P_2 symbolize the projection matrices and N_1 is the nilpotent matrix. A good reference for this computation is Rabenstein's book [4].

The request that a student learn how to compute the Jordan form comes with a warning: If a student is taught this representation in a linear algebra course and the student finishes this part of the course with only an algorithm for making this algebraic representation, then it is a failure. It is the geometry that is important! The geometry! After all, the word projection suggest the geometry.

Asking for the Jordan Canonical Form to be included in the requirements with linear algebra might be a point for discussion. Its computation for text book type problems can be done without much difficulty by hand. For the general (even small) matrix, the computation is notoriously unstable. Yet, I defend this request for that form. I defend it by asserting that an understanding of the representation for a matrix leads to an understanding of how to compute the exponential of a matrix. Just beyond this is some of the beautiful structures in analysis: that of a semigroup of operators and of what subspaces are mapped into themselves by a linear operator. Most important for this discussion, these ideas suggest what parts of an initial value for a differential equation might become insignificant in the solution.

I realize that the notion of a semigroup of operators is a long way from "7-into-4", but it is well for us to look down the road to what generalizes in order to keep focused on the important ideas.

IMPORTANT IDEA II: THE COMPUTATION OF e^{tA}.

It was clear that I was going in the direction of emphasizing the computation of e^{tA}. This solution for systems of equations is readily accessible from the notions I have suggested before. More will be said about computation of this function later. Not only does it solve many of the problems that are typically introduced in a sophomore differential equations course, but also it ties together many of the important ideas associated with constant coefficient nth order equations, or systems.

You know the pattern from here: if A is a 3×3 matrix and is represented as above in Jordan form, then

$$e^{tA} = e^{2t}P_1(I + tN_1) + e^{3t}P_2$$

Whatever is fractious in the computation of e^{tA} is made up by the robustness of the ideas. All the questions about the long range behavior of the solutions of the systems are tied up in the eigenvalues, their exponentials, and the associated projections.

What rich concepts are in this computation!

IMPORTANT IDEA III: CONNECTIONS SHOULD BE MADE BETWEEN THE INTERPLAY OF THE LOCATION OF THE EIGENVALUES, THE ASSOCIATED INVARIANT SUBSPACES, AND THE CHARACTER OF THE SOLUTIONS OF THE EQUATIONS.

The Gerschgorin Circle theorem is not so hard to prove. It gives notions where the locations of eigenvalues are and, hence, understanding of the asymptotic properties of solutions for the differential equations.

Theorem (6CT): Every eigenvalue of the $n \times n$ matrix A lies in one of the (up to n) disks:

$$D_p = \left\{ z : |z - A_{pp}| \le \sum_{j \ne p} |A_{jp}| \right\}$$

This theorem is important because the geometric location of each eigenvalue gives information about which initial values might decay and which persist as the solution of the equation evolves. These ideas are accessible through the Jordan Canonical Form and the Gerschgorin Circle Theorem.

The graph in the picture below represents the ideas that I want. If you were told that this is the trajectory of a solution for a linear, three dimensional differential equation, you would predict that the associated matrix had one real eigenvalue and two pure imaginary eigenvalues.

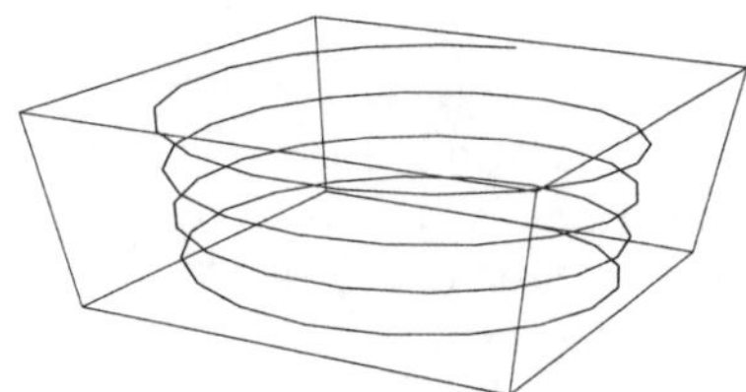

Figure 1: Graph for a solution of a differential system.

IMPORTANT IDEA IV: APPROXIMATIONS OF NON-LINEAR EQUATIONS WITH LINEAR SYSTEMS.

Taylor series have many uses in many places. At no place is the simple first order and second order Taylor series as important as it is in the application to approximating non-linear, differential systems. The notion for vector valued functions that

$$\begin{aligned} f(x) &= f(a) + \langle f'(a)(x-a)\rangle \\ &\quad + \frac{1}{2}\langle x-a, f''(a)(x-a)\rangle \end{aligned}$$

where f is a real valued function of an n-dimensional variable. Here, $f'(a)$ will be an n-dimensional vector and $f''(a)$ is an $n \times n$ matrix.

With this calculus structure, the system

$$\begin{aligned} \frac{dx}{dt} &= y + x^3 \\ \frac{dy}{dt} &= x - y + 2xy \end{aligned}$$

is approximated near $(0,0)$ with the linear equation

$$\begin{aligned} \frac{dx}{dt} &= y \\ \frac{dy}{dt} &= x - y \end{aligned}$$

which has trajectories such as in the Figure 2a. The quadratic approximation near $(0,0)$ would be

$$\begin{aligned} \frac{dx}{dt} &= y \\ \frac{dy}{dt} &= x - y + 2xy \end{aligned}$$

with graph of the trajectories shown in Figure 2b. Finally, the graph of the trajectories for the original cubic equation is shown in Figure 2c.

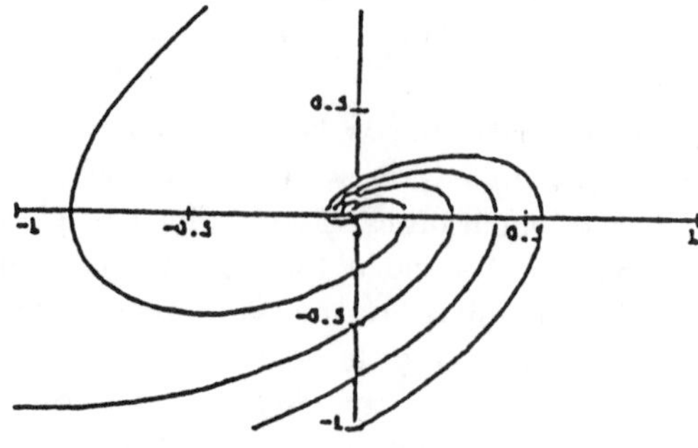

Figure 2a: A linear approximation.

The accuracy of this approximation and the accuracy of the approximation of the solution of the equation is rich with ideas. The rate of growth of the remainder of the Taylor series, the nature of the eigenvalues, and the dimension of the set of equations are considerations for how to predict the accuracy of the long range forecast for solutions.

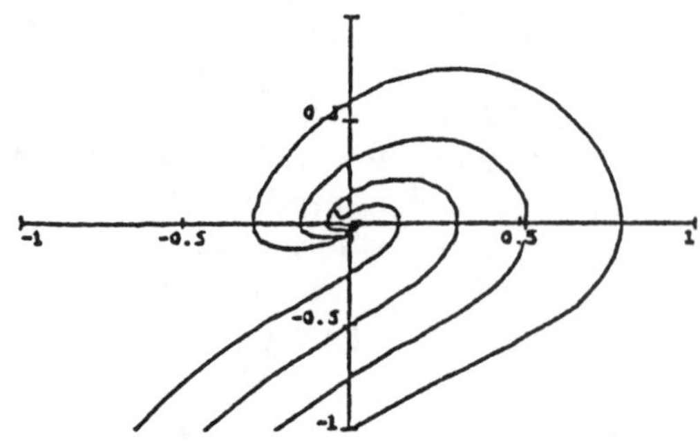

Figure 2b: A quadratic approximation.

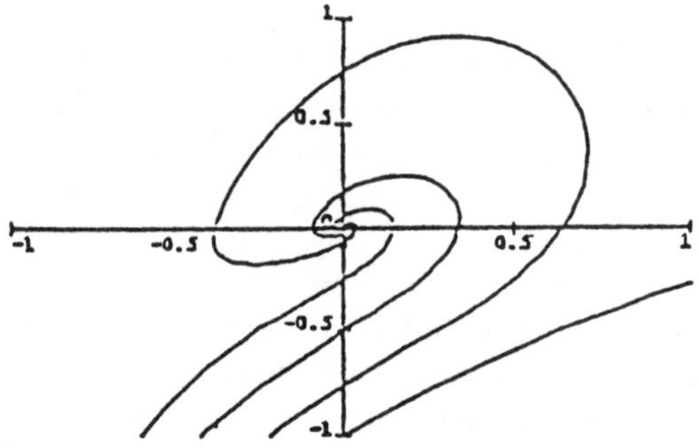

Figure 2c: Graph for a cubic nonlinearity.

IMPORTANT IDEA V: LONG RANGE FORECAST AND STABILITY FOR SOLUTIONS.

As graduate students in various engineering disciplines talk to me about the problems that they are working on, I do not hear them saying that they wish they could get explicit solutions for large systems. What they are interested in is the long range forecast and asymptotic properties of solutions.

As an example, a graduate student from electrical engineering was modeling the electrical grid for the distribution of electrical power in the Western United States. He wanted to make a model for the generation of power to cool Phoenix in the afternoon. A part of the model was to shift power from the network in the Northwest. There is the new problem: what will happen if the proposed change takes place? It is proposed that the nature of the generation of electricity from Glen Canyon Dam be radically changed from the current daily opening and closing of the gates of the dam. This periodic surge, releasing water into the Colorado River, is being questioned by the Department of Interior's Bureau of Reclamation. The recommendations of the draft environmental impact statement are intended to reduce threats to Native American Artifacts, endangered fish species, and recreational facilities in the portion of the Grand Canyon which lies downstream from the Glen Canyon Dam. There is a recommendation to decrease the flow and, in any case, to decrease the "tidal surge" that comes from opening and closing the gates daily.

There will be resulting changes in the electrical grid for the Western United States as more power is shifted from generating plants in the Northwest to cool the Southwest.

The graduate student does not hope to solve the system of equations. He wants to know the character of whatever stationary solutions there might be, and the quality of these solutions as attractors. The shifting of the power will be arranged in accordance with the stability of the resulting model.

It is not the hope that sophomores will finish their second year mathematics knowing either how to obtain analytic solutions for large classes of nonlinear systems or even how to analyze the stability of classes of large nonlinear systems. But, they should have an introduction to these ideas so that as they are encountered again, there will be already a place to incorporate them into their thinking.

IMPORTANT IDEA VI: DISCRETE MODELS AND CONTINUOUS MODELS

There are many different ways to come to an understanding of a model. Two that reflect each other so well are the discrete models and the continuous models. By the continuous models, I mean the analysis of models with the tools of differential equations.

One has only to explore discrete models a little and to note, for example, the summation-by-parts formula:

$$\sum_{j=0}^{n} A_j \Delta(B_j) = (A_{n+1}B_{n+1} - A_0 B_0) - \textstyle\sum_{j=0}^{n} \Delta(A_j) B_{j+1}$$

where $\Delta(B_n) = B_{n+1} - B_n$. On first seeing this formula, one is jolted by the awareness that there is a parallel calculus in the setting of differences and sums with that of differentiation and integration.

Given A, the equation

$$y' = A(y(t)), \text{ with } y(0) = c$$

is probably best rewritten as

$$y(t) = c + \int_0^t A(y(s))\, ds$$

to make the analogy. In the discrete analog, one asks what will be the form of the sequence y if

$$y(n) = c + \sum_{p=1}^{n} \Delta(s) A(y_p)$$

The answer to such a question comes—perhaps not surprisingly—as a product:

$$y(n) = \prod_{p=1}^{n} [1 + \Delta(s)A]\, c$$

Curiously, this is correct, even if A is not linear. To the novice, inexperienced in discrete models, checking what is the "correct way to say" what we all know to be true in the continuous case is an interesting exploration.

Even the models are interesting to formulate. Suppose a drug is taken that is eliminated from the digestive track so that at time T, there is the portion of the original dosage

$$C_0 e^{-kT}$$

left. The constant k we suppose is determined experimentally. Take a dose at time nT. Immediately after the nth dose the amount present will be

$$C_0 \frac{1 - e^{-nkT}}{1 - e^{-kT}}$$

This result can be obtained from a discrete analysis of the system. Questions of how to regulate the dosage to fall in the drug's effective range can now be addressed. I mention this model because the continuous analog was discussed in the Computer Algebra Systems in Education Newsletter. (See [3].) There, the model for this problem was posed as a differential system, keeping track not only of the concentration in the digestive track, but also in the blood stream.

IMPORTANT IDEA VII: SOLUTIONS FOR SPECIAL VARIABLE COEFFICIENTS FOR NONLINEAR EQUATIONS

Most of us can remember worrying as an undergraduate that the course in differential equations might be a collection of techniques that worked for special classes of equations. Perhaps, even, the course seemed to be taught that way!

In fact, there are some special differential equations that are a part of the cultural heritage of a sophomore differential equations course. One might insist that every bright young mathematical science student should have encountered these equations. There are such names as

Exact Equations,
Cauchy-Euler Equations,
Bernoulli Equations,
Riccati Equations,
Lotka-Volterra Equations,
Bessel's Equation.

Some schools provide their students with a listing of special equations and special forms of solutions. We faculty have, instead, an encyclopedia such as Zwillinger's *Handbook of Differential Equations* (5).

No one would say that students should have solutions for these equations as a part of their recall. But the equations are common enough among the mathematical sciences that their solutions should be accessible.

Incorporating Technology into Undergraduate Education

I have asked that many ideas should be conveyed by the end of the sophomore year. Of course, differential equations is but one of the seven to go into the four. Can this be done? One advantage is the attempt to put ideas in place at the appropriate time. There is another advantage that we have as we stand near the midpoint of the last decade of the twentieth century. We have a technology in the form of computer algebra systems that allows directions many of us would have felt unbelievable just a few years ago.

Technology now allows us to solve equations previously inaccessible.

The sophomore differential equations course that you and I studied is much different from the differential equations that should be taught now. For us, it was important to get the forms for equations organized and to develop skills for integrating the appropriate forms. The coefficients for the equations were designed so that the work would be as simple as possible. Perhaps the coefficients were even integers. Yet, there were tricks for setting up the right forms and there were long pages of ingenious and complicated techniques of calculus to apply to these complicated forms. Technology can do that now. While I do not in any way mean that those forms can be ignored by students, what can now be the emphasis is why these equation were studied anyway, what are the implications of the solutions, and how changes in the parameters affect changes in the outcome.

The reason that the emphasis can be changed is the existence of the computer algebra systems that will quickly derive analytic solutions for these equations.

Here is an example: In their book *Differential Equations Laboratory Workbook*, [1] Borrelli, Coleman, and Boyce describe a study of lead intake and excretion of a healthy volunteer living in a lead contaminated area. When lead is ingested, their work models how it moves slowly into the skeletal structure, but also is lost slowly from the bones. One could ask what happens if a person is removed from the lead contaminated environment. How long will it take to have the lead fall to normal levels in soft tissue? The model for the system is

$$\begin{aligned} x' &= -.0361x + .124y + .000035z \\ y' &= .0111x - .0286y \\ z' &= .0039x - .000035z \end{aligned}$$

Here, x represents the level of lead in the blood, y represents the level of lead in other soft tissue, and z represents the level of lead in the bones. It is assumed in the above equations that there is no input from the environment, that the subject has been taken to an uncontaminated environment. The initial distribution of lead in these compartments is not specified for this discussion. If one were to ask how the model predicts the level of lead dissipates, one might compute the exponential of that matrix.

No human would want to do that! Here are the simple commands to compute e^{tA} in MAPLE syntax.

```
*  A:=matrix(3,3,[-.0361,.0124,
       .000035,.0111,-.0286,0,
       .0039,0,-.000035]);
*  exponential (A,t);
```

It's that easy. Solutions and analysis of problems are accessible to undergraduates that were inconceivable before.

Technology allows us to compute and graph numerical solutions on one platform.

I used to derive the standard model for a nonlinear pendulum and then explain that we could not solve the resulting equation or compute the solution numerically, without writing code in FORTRAN, or C, or whatever. Having a numerical solution, we might have to invoke a different computer program to plot the graph. Rather, I continued the lecture, got the linearization of the equation, and drew the solutions for that equation by hand.

Now I give the same lecture, but ask that the students hand in the next week a graph of solutions of the nonlinear pendulum problem and the linearization superimposed. And I expect them to explain what they did in English. The students do this even though they know not one command of any of those above mentioned computer languages. They do it all using only one program, including the write up in English.

The examples were accessible before, but until recently, grasping the visualization of the solutions took several computer platforms: one for computing, one for graphing, one for incorporating into text. Now, with one platform, one can go from the statement of the problem to its numerical calculation to its graphical visualization. The student can stop at any point along the way: numerical scheme, accuracy of computation, accuracy of the approximation to the solution, accuracy of the asymptotic properties predicted by the computed solution, implication of the solution to the place from which the example came.

The computer algebra systems have indeed changed the way we do and teach mathematics. They cannot be ignored in any mathematics, science, or engineering curriculum in 1994.

Technology frees us to see the pattern, instead of requiring just the details.

I will illustrate a use of the technology that allowed understanding beyond the ordinary in the context of a partial differential equation. The equation is

$$2\frac{\partial^2 u}{\partial x^2}+3\frac{\partial^2 u}{\partial x \partial y}+\frac{\partial^2 u}{\partial y^2}=0$$

with

$$u(0,y)=e^{-y^2} \text{ and } u(0,y)=0$$

The solution of the equation is

$$\begin{aligned} u(x,y) &= 2\exp\left(-\left[y-\tfrac{x}{2}\right]^2\right) \\ &\quad -\exp\left(-\left[y-x\right]^2\right) \end{aligned}$$

The students found and reported the solution using a single platform. They also plotted the graph. The plotting was provocative: the solutions are exponentials of negative relations between x and y that should decay as x and y increase. Yet, the graph seems to grow. How can this be? (Has MAPLE made a mistake?) I asked the students to compute $\sup u(x,y)$. The students' surprise that the solution grew happened because of the visualization that the technology allowed. That surprise provoked an understanding in unexpected directions.

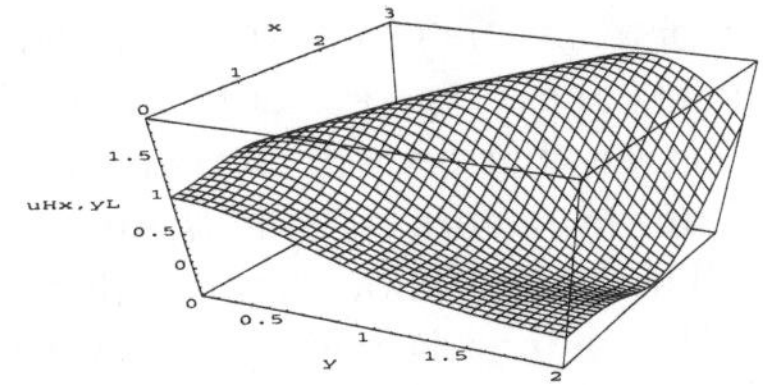

Figure 3: Visualization of a solution surface.

A second example also comes from that partial differential equations class. It is an examination of the equation

$$\frac{\partial u}{\partial x}+2\frac{\partial u}{\partial y}=0$$

with

$$u(0,y)=\begin{cases} 1 \text{ if } y<0 \\ 1-y \text{ if } 0<y<1 \\ 0 \text{ if } 1<y \end{cases}$$

In this example, there is a shock point-a point where characteristic lines cross and there is conflicting information for how the solution should progress. From physical consideration, as formulated in the Rankine-Hugionot Condition, there is a well-defined method to extend the solution past the *shock point.* While it is not so hard to explain the methods in class, it is nearly impossible to draw the solutions at a board with chalk. Yet, students can make the graphs using the technology.

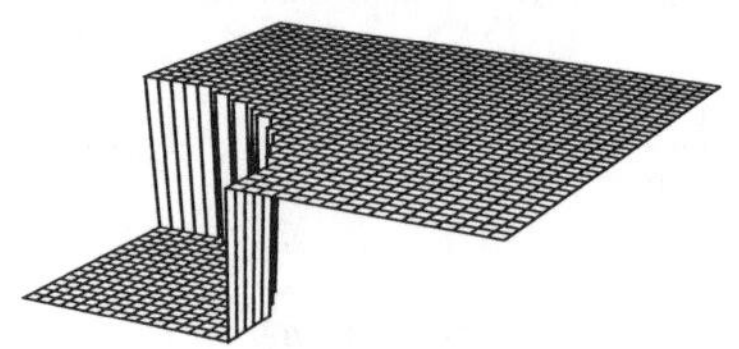

Figure 4: Visualization of a shock point.

I have seen only one undergraduate text that explains this geometric notion in partial differential equations with pictures such as the one that follows. Now, the students can draw the pictures themselves!

Summary

A part of the excitement and challenge of "7-into-4"' is that it begs that ideas be integrated into the structure with an eye on the entire two years, and beyond. It does not take the short view of the current topic, or section, or quarter, or even the year.

What we are here to do is to see the ideas of the four years–and beyond–and to ask how to introduce those ideas that are appropriate for current understanding, are appropriate for preparing patterns, are appropriate for making tools. The students who participate in "7-into-4" must know a collection of standard linear algebra ideas and tools; but this does not mean that we will stop and teach that collection of linear algebra outside the context of the calculus. They must know some collection of ideas about series, but it is not necessary that they stop and study series as a separate three weeks topic. They must know some solutions of difference equations and differential equations, but they do not need to have a quarter devoted to difference equations and a quarter devoted to differential equations.

Students must know how to get solutions for some standard differential equations. Those solutions are accessible already by the use of computer algebra systems. Rather than spending one single class period in working ten variations of a Riccati equation, I would rather spend an hour showing how the equation arises in biology, chemistry, or physics. I would rather show the character of the solutions of the equations. Then, when given a system where the character is the same as the character of the previously discussed equation, the student is likely to be able to solve the equation and determine the nature of the solutions, even though the special techniques of solution may not work.

By integrating the tools of the analysis, the perspective of the first two years of university mathematics as consisting of three semesters of calculus and differential equations may be changed. We believe that the students come out with the perspective that applied mathematics integrates many ideas. This change may equip them to view their other studies differently.

Finally, it does not stop here. Ask the students if they expect to use these ideas during the remainder of their undergraduate studies. Ask them again in an exit interview if they did use the ideas from the first two years of mathematics in their upper division studies. The success of this program is not just in the hands of the mathematics faculty. It is also in the hands of our colleagues in biology, chemistry, social sciences, economics, and other disciplines. They must know what we are doing and are preparing their students to do. Even though we might construct a beautiful program for the first two years of mathematics, it is of little value if the students never use the ideas when they leave our class.

The evaluation of the work that we do at this conference, that we do in our home universities, that we do in our influence on national programs is in the hands of our colleagues and, ultimately, in the work of our students.

References

[1] R. Broiler, C. Coleman, W. Boyce, *Differential Equations Laboratory Workbook*, John Wiley & Sons, Inc., 1992.

[2] Gupta, R.M. Anderson, R. M. May, *Mathematical Models and the Design of Public Health Policy: HIV and Antiviral Therapy.* SIAM Review, Vol 35, No 1(1993), pp 1- 16.

[3] J. V. Herod, *A MAPLE Model for a Pharmacokinetics Problem*, Computer Algebra Systems in Education Newsletter, No 16(1993), pp 1-3.
[4] A.L. Rabenstein, *Elementary Differential Equations With Linear Algebra*, 4th Edition, Harcourt Brace Jovanovich College Publishing, 1992.
[5] D. Zwillinger, *Handbook of Differential Equations*, Academic Press, Inc., 1992.

Response to Differential Equations in the Core I

Donald Bushaw[19]

In the following, I will make some possibly disconnected remarks on the problem of "7-into-4", in particular as it applies to introductory material on differential equations. It is clearly a question of curriculum, a word that has been shown a certain amount of disrespect here today—not unfairly.

"Curriculum" is the Latin word for "race track." I am among those who think that the associations of the idea of a race track are unfortunate here. In education as in many other human endeavors, quality is not closely correlated with velocity. In other words, learning is not well measured by "coverage." Surely the differential equations part of a sequence of courses like the one we've been discussing should emphasize representative approaches to the characteristic problem or problems of the subject, without attempting to say everything, or even everything that is important, about the subject.

What, after all, is the characteristic problem of the subject? Surely it is that of inferring interesting properties of solutions of differential equations from the differential equations themselves. (Solving the equations is only a very special way of doing this, when the equations can be solved at all.) If modeling is involved, then this fundamental process extends to dealing with links with the "real world" at both ends, among other things, and then the problem may become more meaningful or more attractive to students; it may also be considerably more difficult.

There are many ways of attacking the "characteristic problem," and in approximately one seventh of a two-year course it is not possible to come near cataloguing all of them. In his paper, Professor Herod has quite reasonably chosen to emphasize two of them, namely the use of linearity, especially in the form of analyzing a linear approximation, and the use of computers, especially computer graphics. These are powerful methods when they apply, and—as he has shown—they do apply to important and interesting examples.

Of course, there has not been time to present all of the advantages and limitations even of these two strategies.

For example, one advantage of linear systems, and therefore of linear approximations, that Professor Herod has passed over in silence is that it leads by way of e^{tA} to a very powerful matrix version of the method of variation of constants. At the same time, linear approximations do not always make it possible to discover all qualitative properties of the solutions of a differential equation or system of differential equations, such as periodicity or asymptotic behavior not related to equilibrium. Computer graphics lose some of their punch when more than two or three variables are involved, as in a system of three differential equations.

The "important ideas" enumerated in the paper are indeed important, though naturally some of them are probably more important than others.

I think also that Professor Herod's observation that it is now possible to go on one platform through all the steps (statement, numerical solution, visualization, write-up) in studying a differential equation is of great interest and practical importance.

I would have liked to see mention of a few other basic topics, such as theorems on the existence and uniqueness of solutions, which certainly would not need to be proved in this setting, but are easy to state and easy to understand, and help to justify the very language of the subject, as when we speak of the solution that does such and such.

Another possible topic, though it is hardly conventional, is what engineers sometimes call system identification or parameter estimation. In a way, it reverses the usual problem and works its way back from data on one or more solutions of a differential equation supposed to be of a certain type to a specific equation of that type. Following the popular TV show, this might be called the "Jeopardy" approach to differential equations.

A particularly important point of view toward the subject, I believe, is that encapsulated by Poincaré in the titles of several of his papers by the phrase "the functions [he said 'curves'] defined by differential equations." This is another formulation of what I called "the characteristic problem" a few minutes ago. It can be illustrated beautifully by

[19]Washington State University, Pullman, WA 99164

simple examples, e.g., by using the system

$$\begin{aligned}\frac{dx}{dt} &= y\\ \frac{dy}{dt} &= -x\end{aligned}$$

to define the sine and cosine and to obtain as many of their properties as you want, and thereby to establish the whole subject of trigonometry. This is, in microcosm, what the subject of differential equations is all about.

Response to Differential Equations in the Core II: ODEs Renewed

Courtney Coleman[20]

Ordinary differential equations have been in the mathematical curriculum almost as long as calculus, and for good reason. ODEs are one of the premier mathematical tools for modeling dynamic physical phenomena, and that has been true for 300 years. However, what is taught about ODEs and when and how it is done is undergoing a sea change. There are five reasons for this:

- Reform calculus emphasizes applications—and what better source at the introductory level than ODE models?
- The availability of good software and hardware has radically changed what students can do with, and understand about ODEs.
- Interdisciplinary activities with scientists and engineers have a natural focus in ODEs.
- The renewed popularity of discrete dynamical systems at a pre-ODE and even precalculus level has stimulated interest in the ODE syllabus.
- Chaos is here (for how long?), students have heard about it; and we will address it in our ODE and Discrete Dynamics courses, or be charged with ignoring an important contemporary scientific idea.

Here are the features of the new ODEs:

- More on the geometry of solution curves and orbits, and how these geometric features tie in with the form of the ODEs.
- Less drill on finding solution formulas. Most closed form solution formulas are found by fiddling with an ODE until it has the form of the derivative of something, and then the formula appears after an integration. Much of this will be done in the calculus course; computer algebra systems are a part of DERIVE, MAPLE, and MATHEMATICA and widely available for student use.
- More on how solutions and orbits respond to changes in parameters and initial data. Computer solvers and graphics make this easy for students, even when the solution formulas are complicated or non-existent. Sensitivity to data and parameters is a critical part of the study of any ODE model system.
- More on how ODEs respond to pulse, triangular, or square wave inputs. Good solvers have these as predefined functions, and they are crucial in dealing with applications in engineering and science.
- Much more emphasis on graphics calculators and computers as a way to understand and work with ODEs and their solutions.
- More on modeling physical phenomena with ODEs, and then using computer simulations to understand both the ODEs and the phenomena.
- Discrete dynamical models will play an increasingly important role.
- More use of interactive multimedia. The visuals and the applications of ODEs are naturals for this approach.

New ODE books and new editions reflect these trends, and ODE computer laboratory workbooks and ODE solvers are now available. Changes have begun and, I believe, will accelerate.

The USMA, Boston University, Cornell University, Arizona State University, Clemson, Rensselaer, Harvey Mudd, and many other colleges and universities are experimenting with new syllabi for ODEs or discrete dynamics, with ODE computer labs, and with the use of graphics calculators for ODEs. CODEE is a consortium of two and four-year colleges and universities, sponsored by the NSF, that has been running workshops for college faculty on computer experiments in ODEs. CODEE also publishes a newsletter on experiments and new directions in ODEs; write me or send e-mail to codee@hmc.edu to get on the mailing list.

[20]Department of Mathematics, Harvey Mudd College, Claremont, CA 91711

Response to Differential Equations in the Core III: Data Sets in the First Course of Differential Equations

David O. Lomen[21]

While it is impossible to condense 7 courses into 4 without eliminating a considerable amount of material, this exercise will force a strong examination of what is truly important in these courses and eliminate any duplication. It will also allow the design of a set of CORE topics to be covered during the first two years, independent of typical course boundaries, which weave threads throughout the mathematical material. It provides an excellent opportunity to present topics from ordinary differential equations in a manner that illustrates the usefulness of calculus, instead of in an "ad hoc" or "cookbook" manner. It would also permit the intertwining of topics from linear algebra and differential equations in a more meaningful manner than is usually done when linear algebra is given as a prerequisite course. For special situations, such as West Point and some technical schools, it would present a unique opportunity to also coordinate the mathematical material with that of physics, chemistry and engineering in a fully integrated curriculum over the first two years. In that way the mathematics material could be seen in the context where it is needed and some of the artificial boundaries which often exist among separate mathematics courses, as well as among the mathematics, science and engineering courses, could be diminished. This would provide an opportunity for students to learn material in a natural setting, and diminish the opportunity for them to "compartmentalize" this material. While specific constraints may force schools to adopt the "7-into-4" approach, it is not clear that this is the most desirable way for students to learn this material, or even become intimately familiar with the content of this material.

The delivered paper by Jim Herod assumed the availability of a Computer Algebra System for every student, and for such students, the central role of linear algebra and e^{tA} may be appropriate. However, this central role is not appropriate for the many students in large universities who have not had a prior course in linear algebra and do not have access to a Computer Algebra System. The idea of treating higher order equations as systems is great. His added emphasis on geometry is welcome; however, he could go further with phase plane analysis, time series, simulations for applications, etc. His applications to PDE's seem a bit far afield for a first course in ODE's. No mention was made of the use of data sets, and the rest of this response will focus on what could be done with them in a first course in ODE's. My assumption is that students have ready access to user friendly software for graphing and numerical solutions, including operations with data sets (minimally graphing calculators), but not necessarily a computer algebra system.

Students entering a first course in differential equations after a calculus course could benefit from materials of reform projects that have already used the "Rule of Three" where things are considered from graphical, numerical and analytical points of view. For differential equations, the obvious use of the "Rule of Three" would have the graphical aspect be direction fields, phase planes and graphs of solutions, while the numerical aspect would be a numerical solution (analytical needs no elaboration). However, another numerical aspect that should be added (and was also missing from the basic position paper) is data sets. In fact this topic was missing from the papers and discussion on calculus as well. (The first course at West Point involving discrete reasoning and modeling seems to be the appropriate to introduce data, and this use may be easily continued throughout the CORE curriculum.)

Regardless of the form of the activities with data sets, the most significant effect is that the student is more actively involved with the educational process. An additional way of increasing this involvement is have students develop their own; from library sources, doing simple experiments like a tape deck counter or more involved ones using IBM's Personal Science Laboratory (similar to equipment available from Vernier Software). Some of these probes are also available on the TI-82 and TI-85 graphing calculators. These data sets can be used to develop the differential equation, by plotting the numerical derivatives of this data set as a function of the data (or after looking at this graph, some appropriate

[21]Department of Mathematics, University of Arizona, Tucson, AZ 85721

function of the data). Once the differential equation is solved (either numerically or analytically) then the data can be compared with the solution of the differential equation to check your differential equation model. If the course is started by covering direction fields and Euler's method of numerical integration, these techniques are then available as we study the standard analytical techniques, and are used to enhance understanding at each step. In fact these three aspects should become intertwined so as to impart the greatest insight possible.

Students seem to eagerly take to the inclusion of data sets, especially when they collect their own data to analyze in the application sections. We use public domain mathematical software[22] which allows them to quickly edit and analyze the data, as well as develop functions which match the data. They can also easily plot this data on top of graphs of analytical or numerical solutions of an appropriate differential equation. The net effect is to transform the course from one of learning and regurgitating methods of solution to one of thinking, experimenting, and understanding the role of differential equations in explaining dynamic events. It provides an opportunity for students to more fully understand the relationship between the parameters in the differential equation and some process they are observing. Examples of such experiments include: oscillations of a pendulum or spring (like a SLINKY), cooling of coffee, heating of a probe to human body temperature, terminal velocity of an object, rolling cylinders with different contents down an inclined plane, sliding friction, vibration of beams, radioactive decay, and electric circuits.

We look at one example in some detail, namely that of a vibrating spring. Here we fasten a SLINKY to a tripod-board combination and place the distance probe directly underneath on the floor. The spring is moved from its equilibrium position and released, with the resulting motion being recorded as a data set on a computer. The output from a typical experiment is shown in Figure 1.

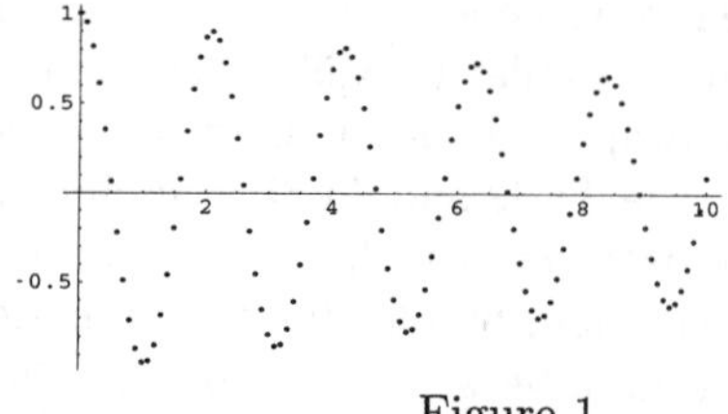

Figure 1

[22]available via ftp from software@math.arizona.edu

While we could simply import this data set into our computer program that gives numerical solutions of differential equations for direct comparison, we encourage students to think about useful operations they could perform on the data before making that step. Since we want to compare with both the numerical and analytical solution of the mathematical model, it is useful to make a change of scale (both horizontal and vertical) so the data starts at a local maximum, or at $y = 0$. This is done very simply and the resulting data is plotted on the graphics screen. At this point (since the mass of the SLINKY, its spring constant, and coefficient of linear damping are unknown) we encourage the students to map out a logical strategy to determine these three parameters. Their usual strategy first determines the maximum amplitude and decay rate (by entering the solution of the differential equation as $ae^{-bx}cos(cx)$, and setting c to zero), before worrying about the oscillations.

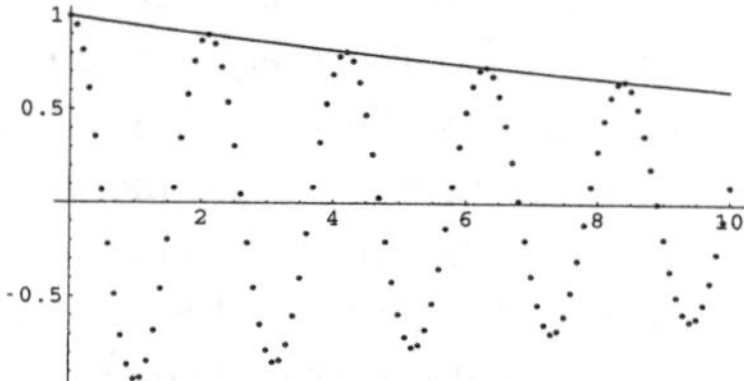

Figure 2

Figure 2 shows the result of this operation, with the parameter c left to be changed so all of the horizontal intercepts for both the data and the solution are identical. (The final comparison is not shown, since the analytical solution completely hides the individual points.) At this point, the student observes the relationship between the a, b, and c in the above equation, and the letters representing mass, damping and spring constant in the differential equation. The relationship between the initial values of the displacement and velocity in the data set, in the numerical solution and the parameters in the analytical solution must be carefully considered in order to make the three results match. The student may also dynamically see the formation of the phase plane from the time dependent solutions for the displacement and velocity.

The main goal is to have it extremely easy for the students to collect the data, so they can claim it as their own, but not be bogged down with many details of setting up equipment. Starting in 1995 the University of Arizona will teach all of our elementary differential equations courses in a room where each student will have a 486 machine at his or her disposal during class and every four machines will share data gathering equipment. The mix between the number of experiments performed during the regular class period or used as homework will be determined as we develop the program. Note that since we use public domain software, students are free to copy to software and complete their assignments using computers located outside the Mathematics Department.

Probability and Statistics in the Core Curriculum

David S. Moore[23]

Introduction

Mathematics is distinguished from most other sciences by a lack of consensus on the content of an introductory overview for potential majors and other serious students. Chemists offer Chemistry 101–102, typically an introduction to the major branches of chemistry. We traditionally offer a year of calculus, followed in the second year by more calculus and perhaps some differential equations and linear algebra. This is hardly a balanced introduction to the nature and variety of mathematics. Attempts at reform have been common, and so has their failure. Table 1 shows one current reform proposal, the contents of a version of "Mathematics 101–102" developed with NSF funding (COMAP, 1997). There is *no* calculus, which is left for the second year. There is also no statistics, though probability does appear.

MATH 101–102?

- CHANGE (sequences, difference equations, series)
- POSITION (vectors, analytic geometry)
- LINEAR ALGEBRA (matrices, eigenstuff, projections)
- COMBINATORICS
- GRAPHS AND ALGORITHMS
- ANALYSIS OF ALGORITHMS (time-complexity)
- LOGIC AND THE DESIGN OF "INTELLIGENT" DEVICES
- CHANCE
- MODERN ALGEBRA (groups, coding theory)

Table 1

The absence of statistics in "Mathematics 101–102" should attract comment. After all, CUPM recommended in 1981 that "other mathematical sciences courses, such as computer science and applied probability and statistics, should be an integral part of the first two years of study." (See Steen 1989, page 5.) This suggestion has generally brought agreement in principle (though little action). I want to offer a partial disagreement in principle. I will argue that, whatever the merits of "Mathematics 101–102," its authors have done the right thing about probability and statistics. They have included the first and omitted the second. To my taste, they would have done well to exercise the same rigor with respect to computer science, in exchange for some continuous mathematics in the first college year. The point is this: a mathematics core ought to display to students the nature and variety of mathematics, including its applicability, but is not the place to develop the principles of related fields. Probability has an important place within mathematics. Statistics does not, and an attempt to include it will be disruptive.

Probability in a mathematics core

Probability has immediate attractions for a mathematics core. Chance phenomena are part of everyday experience and are important in the pure and applied sciences. Probability, the mathematical description of chance, is therefore especially attractive if *mathematical modeling* is one of the principles guiding the core curriculum. This is not true merely because probability models are interesting and have a wide field of application. Most areas of mathematics, when applied to modeling, describe deterministic behavior. It is intellectually stimulating to see how mathematics can also describe chance behavior. The chapter on "Some miscellaneous applications of simple probability" in Noble (1967) remains a good source of simple physical models. Areas such as learning (of rats, alas, not people), genetics, and transmission of rumors or disease are among the biological applications of probability modeling.

Moreover, probability tools that are both *discrete* and *elementary* are powerful enough to be interesting. Conditional probability and tree diagrams for multistage processes, though in simple settings they amount to little more than codified thinking about percentages, allow striking examples. Topics like finite Markov chains are only slightly further afield.

[23]Department of Statistics, Purdue University, West Lafayette, IN 47907

Example 1. ELISA tests are used to screen donated blood for the presence of HIV antibodies. When antibodies are present, ELISA is positive with probability about 0.997 and negative with probability 0.003. When the blood tested lacks HIV antibodies, ELISA gives a positive result with probability about 0.015 and a negative result with probability 0.985. (Because ELISA is designed to keep the AIDS virus out of blood supplies, the higher probability 0.015 of a false positive is acceptable in exchange for the low probability 0.003 of failing to detect contaminated blood. These probabilities depend on the expertise of the particular laboratory doing the test. The values given are based on a large national survey reported in Sloand et al. (1991).)

Now suppose that HIV screening is imposed on a large population, only 1% of which carry the antibody. Figure 1 displays the tree diagram of outcomes. We calculate easily that the probability that a person chosen at random from this population tests positive is 0.0249, and that the conditional probability that a person who tests positive has the antibody is 0.4016. That is, 60% of positive test results are false positives.

Even though ELISA is quite accurate, most positives are false positives when the test is applied to a population in which true positives are rare. Similar results hold for drug screening and lie detectors. Gastwirth (1987) offers a sophisticated treatment.

Probability also illustrates the *interconnections among subfields* that characterize contemporary mathematics. That is true at the advanced level, where probabilistic tools are important in areas such as number theory, PDEs, and harmonic analysis. But interconnections also appear in the first two years of college study, and are important to solving the "so much mathematics, so little time" problem. The use of combinatorics in calculating probabilities in symmetric settings is well known. Overemphasis on combinatorics has traditionally left students mystified by supposedly elementary probability, so I prefer to play down counting unless it will be studied and used elsewhere in the core. Calculations of continuous probabilities and expected values apply integration in ways that emphasize conceptual interpretations: probability is area under a curve, expected value is a weighted average of possible outcomes. A student who sees why expected value as an integral, $E[X] = \int xp(x)dx$, is the continuous analog of the discrete expected value as a sum, $E[X] = \sum x_i p(x_i)$ gains real insight into integration. Even discrete probability leads to the binomial theorem, geometric series, and other mathematical commonplaces. Finite Markov chains use the language and tools of matrix theory. And so on.

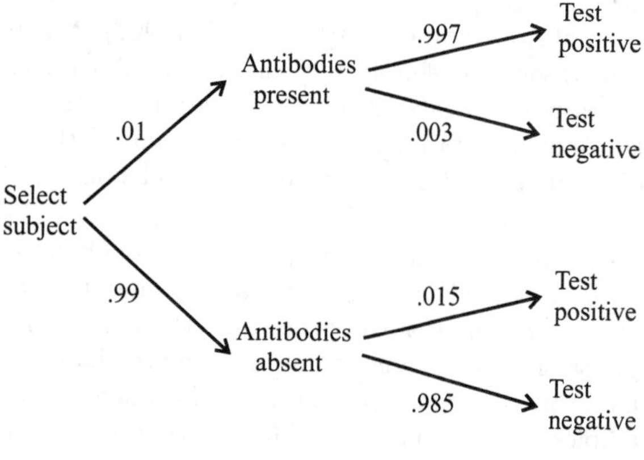

Figure 1

Probability also illustrates the *power of abstraction* in mathematics. The same rules describe all legitimate probability models, though the assignment of specific probabilities may vary greatly in nature and complexity. We can establish many facts once in general, than appeal to them in varied settings. If the use of an axiomatic approach is one of the themes of the core curriculum, the fact that all of general probability emanates from just three axioms is appealing. (I want to demonstrate the power of abstraction much more than I want to work from axioms. And in practice, we can deal axiomatically only with finite probability spaces, lest we meet the ugly fact that we can't assign probabilities to all sets of outcomes.)

Finally, probability lends itself to *computer simulation* as a tool for learning and for modeling. I believe that students ought to meet technology whenever "real" uses of the mathematics would employ technology. Simulation is one such setting. Simulations can demonstrate both the short-run unpredictability of random phenomena and the long-run regularity that probability describes. They allow study of problems that are simply stated but too hard for undergraduate analytical skills. (Probability abounds in such problems. Toss a balanced coin 10 times. What is the probability of getting a run of 3 or more consecutive heads?) Simulation can even help link probability to other topics in mathematics, through for example use of Monte Carlo methods to evaluate definite integrals.

There is room for differences of taste in selecting from this rich array of accessible material. My pedagogical taste runs to the concrete, and to developing mathematics in the context of applications. Simulation appeals to my taste for hands-on work and for technology. I would play down combinatorics and abstract general probability. Whatever our taste, interconnections among core topics are important both for efficiency and to illustrate the unity of mathematics. We should choose topics from probability (and from other subfields as well) with a view to the curriculum as a whole. That is, the first question to ask of any aspect of probability (or linear algebra, or calculus) is not how well it introduces the professional's view of that subfield, but how it contributes to an overview of mathematics in the context of selected aspects from other subfields.

Curriculum planners considering material from probability might seek inspiration from Snell (1988). Though aimed at upperclass students, the book is rich, concrete, makes heavy use of computing, and offers nice historical remarks. It is, however, a mathematician's book with little attention to modeling.

The trouble with probability

The trouble with probability is that it is conceptually the hardest subject in elementary mathematics. Psychologists, beginning with Tversky and Kahneman, have suggest that our intuition of chance profoundly contradicts the laws of probability that describe actual random behavior. They have also demonstrated that incorrect concepts remain firmly embedded in students who can correctly solve formal probability problems. See e.g. Tversky and Kahneman (1983) and the collection by Kapadia and Borovcnik (1991). Garfield and Ahlgren (1988) conclude a review by stating that "teaching a conceptual grasp of probability still appears to be a very difficult task, fraught with ambiguity and illusion."

We run the risk—no, we face the near certainty—that students will learn a formalism not accompanied by a substantial understanding of the behavior that the mathematics describes. Probability is the count of favorable outcomes divided by the count of all outcomes. Probability is area under a curve and can be found by integration. The record suggests that we are unlikely to move most students beyond that level of understanding.

One root of the trouble with probability is lack of experience with the long-term regularity that the mathematics purports to describe. Chance variation is familiar, but chance appears haphazard because we very rarely see the large number of similar trials needed for the emergence of regular patterns. It is not accidental that games of chance, which impose a structure of repeated independent trials, were the historical setting for Pascal and Fermat and have been a staple of teaching ever since. Simulation allows learners to gain some experience with long-run chance behavior. We ought to mix simulation and model building with the mathematics that so strongly appeals to us. We ought to note specific instances (such as the prevalence of runs and other "nonrandom" behavior in short sequences of random trials) in which our intuition fails. But we should also be aware in advance that, given the limited time available in a core curriculum for extended experience with chance behavior, a conceptual understanding of probability will elude many of our students.

The trouble with statistics

The trouble with statistics is that it is not mathematics. It is a discipline that (like economics or physics) makes heavy and essential use of mathematics but has its own subject matter. Many engineers and scientists will find a knowledge of statistics useful and will wish to study the subject. Most mathematics students should study some statistics as a quantitative tool that complements their mathematical training. Selected applications from statistics (or economics, or physics) can add richness to the mathematics core curriculum. But it is unfair to both mathematics and statistics to attempt a substantial treatment of a separate discipline in the mathematics core.

That bald statement reflects the self-understanding of most statisticians. It may surprise some mathematicians, who regard statistics as a (somewhat trivial) field of mathematics. Probabilists, specialists in the field of mathematics most applied in statistics, often know better—note David Aldous's (1994) saying that he "is interested in the applications of probability to all scientific fields *except statistics.*" Let me outline the facts behind the position. Moore (1988) is a more polemical statement of the case.

Statistics is a methodological discipline, the science of inference from empirical data. Under the influence of computing, statistics research and (more slowly) instruction have in recent years returned to their roots in data and scientific inference. Here is the statistician's view of statistics. For more detail, see the essays in Hoaglin and Moore (1992).

Data analysis, the examination o. data for interesting patterns and striking deviatio. from those patterns, is one of the main foci of co. emporary statistics. Data analysis uses both an ample kit of clever tools and a clear strategy for exploring data, but it has no mathematical theory. Graphical display, usually automated and made interactive via software, is always the starting point. Numerical summaries and (sometimes) compact mathematical models follow. Data analysis is specific and concrete. As George Cobb likes to say, "In mathematics, context obscures structure. In data analysis, context provides meaning." In mathematics, abstraction often gets to the heart of the matter. In data analysis, abstraction strips away the details of a particular data set, and so hides the matters of greatest interest.

Designs for data production through sample surveys and experiments have long been a staple of statistics in practice. Their elementary principles are core content in statistics instruction, and their detailed elaborations provide employment for professional statisticians. Although one central idea—the deliberate use of chance selection in producing data—provides a basis for probability analysis, data production like data analysis is not an inherently mathematical subject.

Formal inference is the area of statistics that does have a mathematical theory, based on probability. In fact, inference has several competing theories. The domain of applicability of formal inference is more restricted than that of data analysis. How restricted is disputed. Because statistical inference is a formalization of inductive inference from data to an underlying population or process, it is full of conceptual difficulties and heated debates. The debates concern not the correctness of the mathematics, but the nature and scope of inferential reasoning. Statistical inference is based on mathematical models, but now places heavy emphasis on *diagnostics*, methods that allow data to criticize and even falsify models. The result in practice is a dialog between data and model that reflects the empirical spirit of data analysis. Here is a very brief example of the inadequacy of a mathematics-based approach to formal inference even when diagnostics and philosophy are left aside.

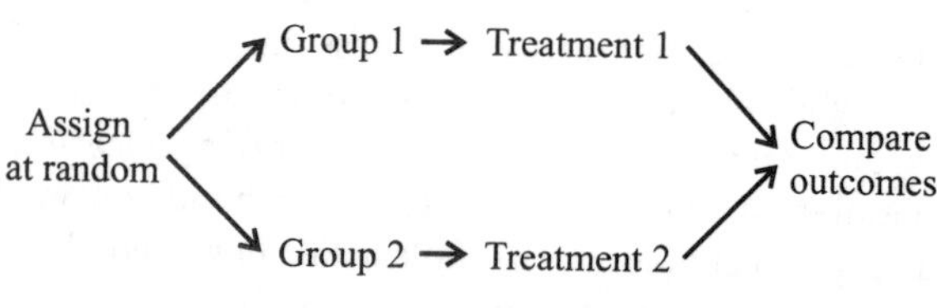

Figure 2

Example 2. A standard setting for elementary inference is the two-sample problem: two independent sets of observations are drawn from populations assumed to be normally distributed. We wish to compare (say) the mean responses μ_1 and μ_2 in the populations. The mathematical model is

$$X_1, X_2, \ldots, X_n \quad \text{iid} \quad N(\mu_1, \sigma_1)$$

$$Y_1, Y_2, \ldots, Y_m \quad \text{iid} \quad N(\mu_2, \sigma_2)$$

Formal inference is based on this model. But the model is radically incomplete. The model, and the formal inference, is the same for two independent samples from two populations and for data from a randomized comparative experiment. Yet the experiment (Figure 2) is intended to allow cause-and-effect conclusions, while an observational study cannot give convincing evidence of causation. The distinction between observation and experiment, and the reasoning of randomized comparative experiments, are among the most important topics in basic statistics. They are inherently statistical, with little mathematical content, and are out of place in a mathematics curriculum.

The American Statistical Association and the MAA have formed a joint committee to discuss the curriculum in elementary statistics. The recommendations of that group reflect the view of statistics just presented. Here are some excerpts (Cobb (1991).

> Almost any course in statistics can be improved by more emphasis on data and concepts, at the expense of less theory and fewer recipes. To the maximum extent feasible, calculations and graphics should be automated.
>
> Any introductory course should take as its main goal helping students to learn the basics of statistical thinking. [These include] the need for data, the importance of data production, the omnipresence of variability, the quantification and explanation of variability.

Data analysis, statistical graphics, data production, and even the somewhat arcane reasoning behind "statistical significance" are mismatched with the mathematical content needed by potential math majors. Yet "statistics" that ignores these topics isn't a responsible introduction to statistics. Statistics in a mathematics core curriculum is an oxymoron.

I should add at once that although mathematics can prosper without statistics, the converse fails. Bullock's (1994) claim that "Many statisticians now insist that their subject is something quite apart from mathematics, so that statistics courses do not require any preparation in mathematics." draws a clearly false implication. Although the place of statistics in mathematics instruction may be marginal, the place of mathematics in statistics instruction remains central.

Applications of Mathematics in Statistics

The decision not to teach statistics for its own sake does not rule out applying mathematics to statistical problems. Consider, for example, the topic of *prediction.*

Example 3. Knowing which of several groups something belongs to can help predict its properties if the groups differ in the property we want to predict. For example, knowing that a hot dog is a "meat hot dog" or a "poultry hot dog" in the government classification helps predict how many calories the hot dog has. Given data on many brands of meat and poultry hot dogs, we find that the mean calorie count is about 160 for meat and 125 for poultry. With no other information, we might use the group mean as a prediction for an individual hot dog.

Interesting use of elementary mathematics arises from looking for *optimal* predictions. The mean is optimal if our criterion is to minimize the sum of the squares of the errors made. The median is optimal if we seek to minimize the sum of the absolute errors. The midrange is optimal if we wish to minimize the maximum error. There is no simple rule for the point that minimizes the median of the absolute or squared errors. This simple setting leads to: the idea of optimization by stated criteria; the fact that the optimal result can vary with the criterion; the fact that the solution may not be unique (the median often isn't) and may not have a simple expression; and of course the technique needed to minimize the criterion functions. Can students show by counterexample that the median, which minimizes the mean absolute error, does *not* minimize the median absolute error? Can they show by example for $n = 3$ that the least median of squares solution is radically unstable?

Example 4. Now suppose that we have more information on which to base a prediction. We have data on an explanatory variable x as well as on the response y we wish to predict. For example, we may want to use the height x from which a rubber ball is dropped to predict its rebound height y.

Plot the data. The graph shows an approximate straight line relationship—not perfect due to measurement error and other factors. If we draw a line through the data, we can use the fitted line to predict y from x. What line shall we draw? Ask the students to discuss criteria. Distances from points to a line are usually measured perpendicular to the line. But in this setting it is usual to use vertical deviations because we are predicting y. The least squares criterion (minimize the sum of the squared vertical deviations) leads by elementary calculus to recipes for the slope and intercept of the optimal line. Formulating the problem requires more thought than solving it. If we prefer to minimize the sum of the absolute vertical deviations, on the other hand, there is no closed-form solution. If we attempt to minimize the median of the absolute deviations, there is no simple recipe for the solution and the computations rapidly become infeasible.

These examples require little background in statistics; even the goal of prediction can be removed if the instructor wishes. They are also chosen to avoid probability, the branch of mathematics most often applied to statistics. For the purposes of a core mathematics curriculum, good applications in another discipline must be comprehensible without much grounding in that discipline. This is as true of applications to statistics as to physics or economics.

References

[1] Aldous, David (1994), "Triangulating the circle, at random," *American Mathematical Monthly*, 101, 223–233. The remark appears in the biographical note accompanying the paper.

[2] Bullock, James O. (1994), "Literacy in the language of mathematics," *American Mathematical Monthly*, 101, 735–743.

[3] Cobb, George (1991), "Teaching statistics: more data, less lecturing," *Amstat News*, December 1991.

[4] COMAP (1997), *Principles and Practice of Mathematics*, New York: Springer. [5] Garfield, J. and Ahlgren, A. (1988), "Difficulties in learning basic concepts in probability and statistics: Implications for research," *Journal for Research in Mathematics Education*, 19, 44–63.

[6] Gastwirth, J. L. (1987), "The statistical precision of medical screening procedures," *Statistical Science*, 2, 213–222.

[7] Hoaglin, David C. and Moore, David S. (eds.) (1992), *Perspectives on Contemporary Statistics*, MAA Notes 21, Washington: MAA.

[8] Kapadia, R. and Borovcnik, M. (eds.) (1991), *Chance Encounters: Probability in Education*, Dordrecht: Kluwer.

[9] Moore, David S. (1988), "Should mathematicians teach statistics" (with discussion), *College Mathematics Journal*, **19**, 3–7.

[10] Noble, Ben (1967), *Applications of Undergraduate Mathematics in Engineering*, New York: Macmillan.

[11] Sloand, E. M. et al. (1991), "HIV testing: state of the art," *Journal of the American Medical Association*, 266, 2861–2866.

[12] Snell, J. Laurie (1988), *Introduction to Probability*, New York: Random House.

[13] Steen, Lynn Arthur, (ed.) 1989, *Reshaping College Mathematics*, Washington: MAA.

[14] Tversky, A. and Kahneman, D. (1983), "Extensional versus intuitive reasoning: the conjunction fallacy in probability judgment," *Psychological Review*, 90, 293–315.

Response to Probability and Statistics in the Core I

Ann E. Watkins[24]

It is hard to disagree with David Moore's central thesis that "Statistics in a mathematics core curriculum is an oxymoron." In the best of all possible worlds, statistics would not be considered mathematics. However, in this world, we must blur the distinction for the benefit of our students. Statistics must be part of the mathematics core because the students who take that core need statistics and there is no other place to get it.

The mathematics core contains the only common courses taken by those majoring in the sciences. Along with mathematics majors, physics, computer science, engineering, chemistry, geology, economics, and even statistics majors take all or most of the first two years of college mathematics. These students need to understand statistics before they get very far into their majors.

How can students do a physics lab with no knowledge of statistics? They can't, so the first "lab" in the lab book used for the first physics course for science, computer science, mathematics, and engineering majors at my university (written to go along with Resnick and Halliday) is actually a crash course in statistical recipes. By page four, the manual is presenting the central limit theorem and confidence intervals. Our students deserve better.

The freshman biology course is also taken by many of the same students who take the mathematics core. One lab from the manual for that course gives three measurements taken on one kind of snail at two different locations. The students are asked to "analyze these data." What help are they given? They are given a computer program that provides what must be mysterious computations.

Now that the first Advanced Placement course and examination in statistics have been offered, many of our math/science/engineering majors will come to the university having already taken a substantial statistics course from a high school mathematics teacher. Statistics will be part of their mathematics core. They will have a tremendous advantage in those science labs over students who didn't have the opportunity to take AP Statistics–if we fail to include statistics in the university mathematics core.

If statistics is taught in the first two years, must it be part of the mathematics core? Moore answers clearly that it should not–and he speaks for many statisticians. He is but the most articulate and the most willing to stick out his neck. However, we have no choice but to teach statistics as part of the mathematics core. Table 1 shows some statistics from the last Conference Board of the Mathematical Sciences (CBMS) survey (Loftsgaarden, Rung, and Watkins, 1997).

The statistics now being taught is taught in mathematics departments because the vast majority of colleges and universities have no statistics department to teach it.

The great service Moore has done in his paper is to remind us that statistics is not mathematics and so if mathematics departments are to teach statistics, they must include data analysis, statistical graphics, data production, and the reasoning of statistical inference. Many mathematicians have already realized that some retraining is necessary, largely thanks to Moore's earlier paper in the *College Mathematics Journal.* The Statistical Thinking and Teaching Statistics (STATS) workshops—which received far more applicants than could be accommodated—and the Activity-Based Statistics materials and workshops are first steps in that direction.

Although most colleges and universities do not have a statistics department, most do have statisticians. A perusal of the *Directory of Members* of the statistical societies confirms that many statisticians are in mathematics departments, where they teach and coordinate the statistics courses. Statisticians or mathematicians, those who teach statistics in mathematics departments by and large teach good statistics. What other explanation could there be for the popularity of the textbook by Moore and McCabe?

[24]Department of Mathematics, California State University - Northridge, Northridge, CA 91330

	Mathematics	Statistics
University (PhD)	169	63
University(MA)	242	8
College (BA)	985	0
Two-Year College	1023	0
Total	2419	71

Number of Mathematics and Statistics Departments in Colleges and Universities

References

[1] Don Loftsgaarden, Don Rung, and Ann E. Watkins. *Statistical Abstract of Undergraduate Programs in the Mathematical Sciences in the United States: Fall 1995 CBMS Survey.* Washington, DC: Mathematical Association of America, 1992.

[2] David S. Moore. "Should mathematicians teach statistics?" (with discussion). *College Mathematics Journal*, 19 (1988) pp. 3–7.

[3] David S. Moore and George P. McCabe. *Introduction to the Practice of Statistics*, 2nd Ed. New York: Freeman, 1993.

Response to Probability and Statistics in the Core II

R. A. Kolb[25]

David Moore presents an excellent, strong argument for the basis and need of probability in the mathematics core. He also mentions some problems with the difficulty of seemingly simple concepts. My own experience parallels Moore's with respect to counting problems. The average student becomes confused and frustrated very quickly when problems get beyond the simple combination and permutation formulas. However, top students are typically fascinated to see how very difficult problems can easily be solved with powerful techniques such as Inclusion/Exclusion. I suggest that basic core courses present only the basic counting rules unless there is a desire to motivate a group of top students to study further in discrete mathematics.

Moore suggests that technology in the classroom be used to automate "rules", but that procedures/formulas/derivations be kept that provide "insight." My example would be the presentation and use of Sums of Squares (SS). I no longer teach the ingenious methods of calculating SS, but I do teach the concept and graphical interpretation of partitioning of SS and require the students, at least one time, to demonstrate to themselves that, with a proper partitioning, cross-product terms sum to zero.

Three of Moore's stronger statements concerning statistics are:

1. "Statistics does not (belong in the mathematics core) and an attempt to include it would be disruptive."
2. "The trouble with statistics is that it is not mathematics."
3. "Statistics in the Core is an oxymoron."

I would retort in turn:

1. The 1981 CUPM Guidelines (Moore was a member of the statistics subpanel) actually suggests that the core contain two courses, one in probability and one in statistics. The single course that the guidelines present contains a significant amount of statistics. The problem is more with time than with a pedagogical argument of whether or not statistics belongs in the core. Those of us who have taught courses with both probability and statistics always feel that more time is needed in both areas to properly develop. It would seem clear to me that mathematicians need statistics. For example, they frequently have to collect data in order to empirically verify their own models. Also, they often work with stochastic variables.
2. Maybe more to the point here is that:
   - "The trouble with statistics may be that our textbooks (courses) are not written (taught) as statisticians practice." This is echoed in the latest Preface to the CUPM Guidelines.
   - "The trouble with statistics in mathematics programs is that it is not taught the way mathematicians will use it in practice."

   We have heard similar statements before with our mathematics books and courses. As Alan Tucker had said, most of our students will graduate only with bachelor degrees and move directly into the work force. We owe it to them to show them how to apply mathematics to real problems. This often includes use of statistics.
3. Is this really a pedagogical argument or are we talking about something else. Maybe the concern is more with protecting "turf." I am as concerned about that at my school as anyone anywhere else. Large schools with statistics departments might be better served by requiring their mathematics students to take a separate statistics course from that department later in their program. At my school, statistics is naturally placed in our mathematical science department. I would argue that "Statistics" as a discipline is naturally imbedded in Mathematical Science.

Should statistics be in the core? After saying all of this, if I had to drop one subject out of the first four courses, it would be statistics. But, who defined that the core can only be four courses? Would we really want to graduate a mathematics major

[25]United States Military Academy, West Point, NY 10996

without a course in statistics. "Let's teach what mathematicians need to practice mathematics. If they really don't need statistics, then throw it out." Of course, I think otherwise.

Probability and Statistics in the Core III: Is It Really Such A Tight Squeeze?

Sheldon P. Gordon[26]

While statistics has a very different philosophy from mathematics, as David Moore points out, that doesn't necessarily mean that statistics courses can, or should, be left exclusively as the province of statisticians. The faculty in engineering departments who offer statistics courses may have no more formal training in statistics than math faculty do; what is important is that whoever teaches such courses (statisticians, mathematicians, engineers, psychologists, ...) have some experience with using statistical ideas and methods to study real data and have the ability and enthusiasm to communicate this to their students. An instructor with a background in theoretical mathematics or theoretical physics or engineering is not necessarily incapable of teaching such a course effectively; the danger is that he or she will present the statistics from the point of view of theoretical mathematics, physics or engineering. In large measure, then, this becomes a matter of training and re-education. The MAA, through its committee on statistical education, has received large-scale funding from the NSF to conduct a series of summer workshops to train mathematics faculty to teach statistics from a modern, applied point of view. The Statistical Thinking and Teaching Statistics Project is headed by George Cobb of Mount Holyoke College and Mary Parker of Austin Community College.

The question we face here is not one of a turf battle between mathematicians and statisticians. Rather:

> At a minimal level, how do we expose students in mathematics intensive fields to some fundamental statistical and probabilistic ideas in the time frame that typical students have available during their first two years of undergraduate study?
>
> At a maximal level, how do we fit all of the statistics and probability needed by these students into those first two years?

The problem has become exacerbated with the recent ABET (Accreditation Board for Engineering and Technology) decision to require a course in probability and statistics for all engineering students. Many schools have now been forced to come to grips with how to accommodate this in an otherwise overfilled curriculum. Some schools have already done so with seemingly little difficulty – the engineering department offers its own engineering statistics course. Thus, the seven into four problem actually becomes a six into four problem.

Mathematics departments must also come to grips with this problem. Otherwise, by a simple exercise in induction (let computer science departments teach discrete mathematics, let engineering or physics departments offer differential equations, ...), the entire problem can disappear. Mathematics departments will be left teaching only remedial algebra courses. However, this is not an acceptable solution for any of us.

If we are to resolve this problem, we have to consider three separate components:

- Paring significant amounts of material from existing courses;
- Integrating material from different courses in new ways;
- Preparing students differently for these courses.

Lessons from the Calculus Reform Movement

According to a recent study conducted by the MAA, about 56% of all colleges report some level of implementation of calculus reform. About 600 institutions are currently using materials from the major calculus reform projects; many others have developed their own materials. Many of the instructors implementing these reform courses report that they cannot conceive of going back to teach traditional calculus courses. In turn, these efforts have spawned a variety of related projects to reform other courses in the mathematics curriculum, most notably in multivariable calculus and differential equations to reflect the different calculus experience students are receiving and in precalculus

[26]Suffolk Community College, Selden, NY 11784

to reflect the different type of preparation that students now need for calculus. Thus, the calculus reform movement has progressed from an innovative experiment to a major change in the curriculum.

This is quite a remarkable achievement, considering that the central mathematics curriculum has been virtually unchanged throughout most of our professional careers. True, Kemeny, Snell and Thompson introduced finite mathematics about 40 years ago, but those courses are still offered primarily to business majors; discrete mathematics has never become quite as important as many of its proponents claimed 10 years ago; statistics has become one of the most important applied offerings in many mathematics departments, but has had relatively little impact on the central mathematics curriculum.

The success of the calculus reform movement, however, proves that dramatic change is possible in the mathematics curriculum. More importantly, it demonstrates that the change can, and must, be accompanied by a paring away of topics that once were considered essential to that curriculum. Students can function in calculus and other courses without secants and cosecants; traditional related rate problems can be removed entirely and no one apparently misses them; technology can provide effective (in fact, preferable) alternatives to the full array of techniques of integration provided students learn to understand what they are doing and develop the ability to select the appropriate tool; mathematicians can survive without proving the mean value theorem after one month of Calculus I.

The key lesson from this is that material can be removed from the curriculum, and it is this message that is essential if one is to have any hope of resolving the seven-into-four problem. And that resolution is going to be far more painful than anything that we have gone through yet, because far more will have to be removed.

There is another significant aspect of the calculus reform movement that has importance here. Rather than viewing calculus as developing the skills needed for subsequent differential equations courses, most of the major projects have moved the study of differential equations into a far more important and central position within calculus itself. By incorporating substantial parts of the subject, particularly from both modeling and qualitative points of view, these projects have completely undermined traditional cookbook-style differential equations courses. It is no wonder that people who have offered such calculus courses are now intent on developing alternatives to the standard differential equations courses so that they can build on the new perspectives and approaches introduced in calculus.

Finally, the calculus reform movement has lead to a fragmentation of the one-time monolithic structure of freshman calculus. New courses have been developed for different institutional settings and different student audiences. With the universal availability of desktop publishing, we should anticipate a continued fragmentation of the curriculum; it is very unlikely that we will return to a single curriculum in the foreseeable future.

Integrating Mathematical Themes

Let's turn now to some comparable ideas regarding the role of probability and statistics in an integrated curriculum. At least one of the major calculus reform projects has incorporated a limited amount of concepts from these areas into calculus. The Harvard materials contain sections on probability distributions as part of a chapter on applications of the definite integral. In our multivariable calculus materials, which are currently in the development stage, we introduce the notions of least squares and linear and nonlinear curve fitting to sets of data. We also have a brief introduction to the idea of using Monte Carlo simulations to estimate the values of multiple integrals.

This is admittedly but a small step in the direction of integrating probability and statistics into a Core Mathematics curriculum. What more can be done? Let me suggest a few specific ideas. In my Calculus I class this semester, I was analyzing some one- or two- parameter families of functions. One was the family

$$f(x) = e^{-(x-a)^2/b}$$

Having demonstrated that each curve is centered at $x = a$ and has points of inflection at $x = a \pm \sqrt{b/2}$, I went off on an unplanned excursion to relate these ideas to the mean and standard deviation of a set of data, introduced the empirical rule for normal populations, and gave a variety of "statistical" applications to reinforce some of the mathematics. With a little planning, this 20 minute excursion could have been easily extended to incorporate more statistical ideas and some examination of real data sets that follow a normal pattern. Early in Calculus II, when we consider probability distributions as an application of the definite integral, I will come back to the normal distribution and consider probability problems associated with it in considerably more depth.

There are many other opportunities is such a course to integrate some probabilistic and statistical ideas. For instance, when introducing the notion of slope of a tangent line, each student can be given a graph of a function with a tangent line drawn at a point. The students would be asked to estimate the slope of that tangent line, the results collected and analyzed as a set of data. Outliers would likely indicate either gross errors in measurement or, more likely, clear errors of interpretation. The data could then be used to estimate a more accurate value for the slope using the mean and/or median. The question of increasing the accuracy by using larger sample sizes could be raised. A comparable exploration could be conducted when the definite integral is introduced by giving each student the graph of a function drawn over a grid and the question of estimating the area is raised.

Similarly, it would be fairly easy to introduce the use of Monte Carlo simulations via computers or calculators in a variety of ways to provide information on different processes, including evaluation of definite integrals. This could serve as the basis for further discussions on probabilistic reasoning. We can discuss the accuracy of the results, particularly as a function of the sample size. We could develop some ideas on confidence intervals for estimating the value of π or a variety of definite integrals or the average value of a function based on simulations. We could look at Riemann sums for a given function using random points generated in subintervals of a random partition of a given interval. In a truly innovative course designed to integrate the statistics and the calculus, we could conceivably build on these examples to consider more traditional statistical problems on estimation. (Admittedly, introducing hypothesis testing would likely be more of a stretch, but it would not be impossible.) On the other hand, it would certainly be simple to extend a brief introduction of least-squares analysis into a relatively full-blown treatment of nonlinear curve fitting based on the examination of data to find trends and patterns. The effects and treatment of outliers could also be brought in at this point.

The key to developing such an integrated course is making some very hard decisions: what topics—both from calculus and statistics—should be left out. It is impossible to succeed in such an effort without very significant cuts from both sides. We need people from mathematics, from statistics, and from the various client disciplines to sit down together and decide which ideas and methods are so central that they cannot conceivably be omitted and which are merely only terribly important. It will certainly not be an easy process.

Preparation for the New Core Curriculum

Despite the most draconian paring conceivable, it is virtually impossible to believe that the material remaining will fit into four semesters. Thus, it becomes critical to find ways to extend the four semesters available, and this extension will have to take place in the courses that precede the first two years of college mathematics. The calculus reform movement has de-emphasized the role of manipulative methods and balanced it with a greater emphasis on conceptual approaches using geometric and numerical methods. As a consequence, much of the time spent on developing high levels of algebraic skill in precursor courses, both in high school and in college, can now be redirected to developing other skills. In large measure, this is the thrust of the NCTM's Standards – less emphasis on routine manipulation and more on conceptual, graphical, numerical and practical approaches. The colleges and universities are just beginning to see the fruits of the secondary school labors as more and more students move on from such high school courses.

Further, the success of the calculus reform movement is leading to a similar revision of the college-level courses that lead to calculus, courses in precalculus, in college algebra and trigonometry, and in elementary functions. AMATYC has developed a set of curriculum Standards that call for the wide-scale development and implementation of these ideas. Such courses will similarly devote less time and effort to developing high levels of traditional algebraic skills than their more traditional versions; instead, more emphasis will be placed on conceptual ideas, applications, and perhaps most importantly from our current viewpoint, on different mathematical content.

Another major thrust in the NCTM Standards is to include exposure to statistical ideas and reasoning from elementary school up through the end of high school. Students are expected to become familiar with looking at data, at observing patterns, at constructing and testing conjectures based on the data, and interpreting their results. The American Statistical Association is assisting in the implementation of this effort by offering teacher training workshops through its Quantitative Literacy Program. Thus, we can look forward to the day when the overwhelming majority of students coming in to college mathematics courses will have had such experiences. The kinds of Core courses that may emerge from this conference should certainly be able to build on this.

In addition, there are a variety of other college-level projects underway that have related goals. For one, I am currently serving as project director of the Math Modeling/PreCalculus Reform Project which has developed a very different alternative to traditional precalculus courses. The mathematical ideas are all developed in the context of mathematical applications with the same spirit and philosophy that pervade all the reform calculus projects. Several of the major themes that are interwoven throughout the Project materials – data analysis and curve fitting, probability, difference equation models, and linear algebra – are quite relevant to the issue here.

The notion of function is totally entwined with data from the very first introduction of the function concept. After a development of the properties and applications of the most basic families of functions—linear, exponential, power and logarithmic—we turn to a full treatment of fitting functions to data to reinforce the behavioral characteristics of each type of function as well as its algebraic properties. We informally introduce least squares as the means for obtaining the best linear fit to a set of data. We introduce the correlation coefficient as a measure of the degree of linear relationship. We then discuss nonlinear curve fitting, have the students perform the appropriate transformations on data sets to linearize them, have the students obtain the equation of the least squares line using a calculator and/or a computer, have them undo the transformations by hand to practice the desired algebraic properties, and then ask them to interpret the results.

Subsequently, throughout the course, as other families of functions are introduced, such as polynomial functions, logistic functions or trig functions, we return to the notion of fitting such functions to appropriate sets of data. At each point, we discuss problems with extrapolating well beyond the set of data, and the uncertainty of such predictions. We also discuss the idea of critical values for the correlation coefficient and how it depends on the sample size. No, this is not intended as a statistics course, but we certainly feel that appropriate ideas from statistics can and should contribute to an effective preparation for calculus.

We also weave ideas on probability, particularly the use of Monte Carlo simulations to give information on various processes, into the entire course. For instance, we look at a Monte Carlo simulation of radioactive decay to give a non-deterministic view to what is actually a random phenomenon and demonstrate how closely the formal treatment with exponential functions mirrors that random process. We not only consider the nature of the roots of a quadratic equation, but also use Monte Carlo simulations to provide insight into the likelihood that a quadratic, or a cubic, has complex roots. We develop the binomial expansion in the context of binomial probability; a standard type of precalculus problem asks: What is the 10th term in the expansion of $(x + y)^{25}$? A typical student's reaction is: Who cares! On the other hand, using the standard test for ESP with a deck of 25 cards consisting of five cards with each of five symbols, we can ask: What is the probability of obtaining 10 right answers? Naturally, the students are far more interested in the answer to this question.

We also consider a variety of problems involving geometric probability to prepare students for scenarios they will encounter in calculus. We introduce the question of estimating the area of a plane region using a Monte Carlo simulation. We similarly introduce the question of estimating the average value of a continuous function using a Monte Carlo simulation.

We even connect the probability explorations and the data analysis theme by generating sets of random data and then finding the best fit to them from among the families of functions considered. For instance, by looking at a Monte Carlo simulation for the area under a curve such as $y = \sqrt{x}$ from 0 to 1, from 0 to 2, ..., we obtain a set of data and the best fit to this set of estimates for the area is given by a power function such as $f(x) = .6667x^{1.497}$ with a correlation coefficient of r = .99998. Thus, students can see how one might come to expect that a formula can provide information on the area of a plane region.

The course is not intended as a probability course. There are no standard balls-in-urns problems or situations that obviously violate students' intuitive beliefs. Rather, probability is introduced in the service of preparing the students for calculus while giving them a notion of the concepts and uses of probability.

Another major focus of our course is a study of difference equations and their applications. This includes first and second order difference equations, both homogeneous and non-homogeneous. In fact, most of the important models that one typically constructs with differential equations, such as population growth, inhibited growth, level of drug dosage in the bloodstream, radioactive decay, Newton's Laws of Motion and projectile motion in the plane, simple and damped harmonic motion for a spring, and the predator-prey model are treated using difference equations. Students become used to thinking about rates of change, the relationship between the rate of change and the quantity itself, the behavior of the solutions, their dependence on initial conditions, and how to interpret the solutions in terms of the processes being modeled.

We also have a unit on matrix algebra and its applications much in the spirit of what one would do in a finite mathematics course, not just as a gimmick for solving systems of linear equations. For instance, we consider models such as Markov chains and geometric transformations, as well as matrix growth models leading to the eigenvalue problem.

Picture a student coming out of such a course, either in high school or in college, and starting the kind of Core Mathematics sequence that might be the solution to the seven-into-four problem. I'm not sure that "problem" is really the appropriate word; rather, I see it as an exciting prospect to design and develop those new courses which will build on the students' experiences and will combine those seven semesters in four.

Acknowledgement

The work described in this article was supported by National Science Foundation grants #USE-89-53923 for the Harvard Calculus Reform Project and USE-91-50440 and #DUE-9254085 for the Math Modeling/PreCalculus Reform Project. However, the views expressed are not necessarily those of either the Foundation or the projects.

Evaluation of Mathematics Core Curriculum Conference April 22–24, 1994

Robert Orrill[27]

Current efforts to bring about mathematics reform in postsecondary education reach back at least to the early years of the last decade. At that time, the signal to mathematics departments of the need to rethink collegiate mathematics was clearly sounded by the Mathematical Association of America's (MAA) Committee on the Undergraduate Program in Mathematics (CUPM) with the publication of its *Recommendations for a General Mathematical Sciences Program.* Though advocating incremental change, this 1981 report forcefully called for the mathematics community to begin reshaping the mathematics major so that it prepared students for a broader range of purposes and professional destinations than had previously been presupposed. Up to that point, thought about improvement in the undergraduate mathematics major had been largely dominated by the perceived need to strengthen preparation for students intending graduate study in mathematics. This perspective tended to place a heavy emphasis on the importance of classical theoretical concerns. CUPM pointed out, however, that diverse new needs for quantitative competence had been created by the burgeoning field of computer science and the emergence of a rapidly-growing number of other professional environments in which sophisticated mathematical understanding was needed to deal with problems of "organized complexity." Preparing students for these new conditions would necessarily involve finding common ground between theory and application, abstraction and practicality. The probable price of not accepting this challenge, the report added, would be to see other quantitatively oriented fields (e.g., the "decision sciences") begin to devise their own means for providing students with appropriate preparation in mathematics.

Since the early 1980's, mathematics reform in collegiate education has proceeded largely within the framework of recommendations and issues presented in the 1981 CUPM report. It has, however, tended to focus on the shape and substance of individual courses (e.g., calculus, differential equations) rather than on the mathematics curriculum as a whole. Therefore, it came as a welcome event when a conference addressing the core curriculum in collegiate mathematics was convened at the United States Military Academy on April 22-24, 1994. The group assembled for three days of concentrated discussion included upwards of 60 mathematicians and educators drawn from upper secondary schools, community colleges, liberal arts institutions, and major research universities. This range of participation was invaluable given that issues surrounding core curricula must necessarily be considered in the context of the exceptional diversity of mission and purpose that characterizes American postsecondary institutions. Such a mix also enabled conference participants to raise, if not answer, many important questions about articulation and cooperation among different educational sectors.

Given this background, it was appropriate and useful that the conference began by taking account of the main reform trends in collegiate mathematics education since the issuance of the CUPM report of 1981. This discussion was keynoted by Lynn Steen, Executive Director of the Mathematical Sciences Education Board, who pointed out the continuity of concerns that has characterized the reform impulse in the last fifteen to twenty year period. In Steen's view, consensus in the mathematics community about the foremost curriculum issues needing resolution has grown to the point in this period that one can discern at least the potential for agreement needed to develop a common plan for the revitalization of mathematics education. Steen's guarded optimism was shared by most other conference participants, who felt that the task facing the mathematics community was less to create new vision and more to make actual changes in practice.

Perhaps the most radical departure of the reform movement is its conviction that shaping "the curriculum in the mathematics major should be shared among the various intellectual and social constituencies of mathematics" (Tucker, 1989). In short, the making of mathematics curriculum should no longer be inward looking and designed primarily to repro-

[27]Executive Director, Office of Academic Affairs, The College Board, 45 Columbus Avenue, New York, NY 10023

duce a mathematics professorate. Rather, it must take into account and be responsive to enlarged and diverse needs for quantitative competence in social, intellectual, and technical worlds external to the mathematics community per se. Probably, the foremost sign that collegiate mathematicians are prepared to pursue this call for greater diversity in academic programs has been the widespread adoption of the 1981 CUPM recommendation "that the undergraduate major offered by a mathematics department at most American colleges and universities should be called a Mathematical Sciences major." (CUPM, 1981). While some consider this change as more symbolic than real in many institutions, its import as a statement of orientation and intent continues to be very important.

Of course, with this call for more diversity and flexibility also comes concern for the coherence and thoroughness of preparation provided to students in the mathematics major. Can the mathematics major serve many masters or only one? How can teaching resources be found and allocated by mathematics departments that is commensurate with this enlargement of purpose? John Dossey, Visiting Professor of Mathematics at the Military Academy, articulated these and other related questions at the outset in setting the stage for the working sessions of the conference. Very usefully, Professor Dossey restated the major issues that he had posed to participants in a communication prior to the conference. The essential questions, he proposed, were as follows:

- Can the mathematics major be integrated through the development of a core curriculum common for all students enrolled in the major given the increasing diversity of their needs?
- How should the common and integrative elements of a core curriculum be defined?
- How can the current seven topics commonly required of mathematics majors be feasibly integrated into a shorter, integrated program of study?
- How does the concept of a core curriculum help clarify and support liberal educational aims that exceed the needs of more narrow technical training?

For the most part, Professor Dossey's questions shaped deliberations throughout the working sessions of the conference. These sessions alternated between plenary assemblies and intervals of small group work. The plenary sessions provided separate attention to the different mathematical sciences courses often required of a student majoring in mathematics – discrete mathematics, linear algebra, calculus, differential equations, probability and statistics. In a complementary manner, the small group work was centered on developing models and approaches for integrating these topics into a core mathematics curriculum. In the main, these working sessions were organized in such a way that conference participants from similar kinds of institutions were grouped together. This arrangement contributed to the efficiency of discussion in what, overall, was a substantively rich, but ambitious, agenda.

Each plenary session focusing on an individual mathematical topic began with the presentation of a paper (prepared and circulated in advance) which was followed by a panel discussion. Perhaps inevitably, much discussion in these sessions concerned whether or not the topic under consideration warranted inclusion in a core mathematics curriculum. Not surprisingly, the majority view seemed to be that all did - though not necessarily in their current formats. The one exception and source of most debate was statistics, with regard to which Professor David Moore of Purdue University took a contrary position in a paper prepared for the conference. Moore, of course, did not question the importance of statistics as a subject but only its inclusion in a mathematics core. In this regard, Moore argued, the trouble with statistics "is that it is not mathematics." Rather, he said, it "is a discipline (like economics or physics) that makes heavy and essential use of mathematics but has its own subject matter." As time in this session ran out, debate about this matter continued in full force.

In this brief report, it is not possible to do justice to the many important points made in the plenary sessions. Indeed, the sessions themselves sometimes did not permit the full pursuit of interesting issues because of constraints imposed by time pressures. Some participants, for example, pointed out that, unlike many other subject areas, mathematics does not have an "overview" course designed to present students with an integrated picture of the discipline. The possible implications of this fact, they felt, should be investigated in any attempt to model a core curriculum. Many others noted that questions about the core content of mathematics could not be disengaged from the processes of teaching and learning employed in mathematics education.

On this score, it was acknowledged that mathematics reform in K-12 education is probably more articulate and advanced than in the collegiate sector. As will be noted later, the desirability of adopting a perspective on mathematics reform that encompasses a K-16 educational continuum emerged as an important emphasis in the conference deliberations.

The work of the small groups much enriched the conference deliberations. In these sessions, there was opportunity to give close attention to issues that could be touched upon only briefly in the larger plenary sessions. Moreover, this was the context in which conference participants most directly undertook the challenge of debating and forming models for integrative core curricula. Not surprisingly, the degree of consensus reached about outcomes varied somewhat across groups. Overall, groups reflecting the circumstances of secondary schools, community colleges, and four year liberal arts institutions, more readily arrived at agreement than did those constituted of participants from large research universities. Generally speaking, however, all of the groups confirmed the importance to mathematics reform of the questions raised by John Dossey. Moreover, most groups agreed that an integrated curriculum could and should be achieved through clarification of the "threads" that connect and unite student learning experiences in all of the core courses. Such threads include mathematical reasoning, problem-solving, modeling, communication, technological competence, data analysis, and the understanding of mathematics as a historical achievement. The majority of participants appeared to conclude that settling on these threads and giving them prominence as organizing principles for course work should be the first priority in developing core curriculum models.

The key to an integrated mathematics curriculum, then, is not to be found in a consideration of separate topics but rather in a determination of those "threads" that should emerge in and weave together all of the core courses. In this approach, the content of any course is not shaped by presupposing a particular professional destination for all mathematics majors, but rather is guided by the aim of helping students achieve a capacity for "mathematical power" that will serve them well in a wide range of quantitatively complex environments. This, of course, suggests that the core must be further unified by creation of a framework and set of criteria for measuring student growth in terms of the common "threads." Such an approach to assessment should de-emphasize sorting and slotting students according to assumed destinations and, instead, encourage attention to student growth in quantitative competence.

Throughout their work, conference participants were stimulated by having in play a working model of a core mathematics curriculum developed by the Mathematical Sciences Department of the United States Military Academy. This curriculum, known as CORE, has been under development since the late 1980's and features the conversion of the usual seven required courses of the mathematics major into a coordinated two year sequence of four courses. This sequence includes special attention to such unifying threads as mathematical reasoning, modeling, scientific computing, writing, and the history of mathematics. The CORE model was thoroughly described for conference participants in an evening session and an explanation of its role in the special mission of the Academy was provided. Although this model must be regarded as something of a special case given the Academy's unique control over student academic programs, the conference participants were impressed by CORE and found its underlying conceptual and organizing principles directly relevant to core curriculum planning at other institutions.

As stated, a major purpose of the conference was to enlarge the perspective on mathematics reform in postsecondary education, to take discussion beyond the reshaping of separate courses such as calculus and to inquire into the coherence and integration of the mathematics major as a whole. With regard to this aim, the conference fully succeeded in at least broaching questions often little addressed by the mathematics community in any sustained or cross-institutional manner. Indeed, the conference also succeeded in opening the reform discussion to considerations greater than the planners may have anticipated or thought possible. In particular, Jack Price, President of the National Council of Teachers of Mathematics (NCTM) called for much closer cooperation between K-12 teachers and collegiate faculty in aligning mathematics reform in secondary and postsecondary education. Earlier in the conference, secondary school participants raised questions about whether student growth models for the core curriculum could work well if they did not take account of the mathematics preparation of high school students. The response in general seemed to be that they could not, and that the ultimate success of the improvements considered at this conference depended upon simultaneous and aligned reform efforts in both schools and colleges.

Finally, some conference participants called attention to the fact that reform still seldom addresses the perplexing issue of what relation courses in the mathematics major should have to improving the quantitative literacy of *all* students. According to one estimate, the core models considered at the conference would reach (even in a modest way) only about 15% of undergraduates in a given institution. Whether this figure is accurate or not, it does seem to be the case that we are still far from being able to define the role of mathematics in any undergraduate curriculum that is genuinely liberal in its aims and does not assume a relatively high degree of specialized purpose. This, of course, is a problem shared in some degree by all subject areas, and answers are to be found only in cross-disciplinary contexts now largely unavailable to the academic community.

Section III
Responding to the Core Curriculum

A Curriculum Reform Workshop/Retreat

Don Small[28]

This Workshop/Retreat was Part II of the "Core Curriculum Conference in Mathematics" hosted by the U.S. Military Academy in April, 1994. The purpose was to facilitate curriculum reform arising from the 1994 Conference. The academic programs of the schools invited represented the broad diversity seen in the colleges and universities of this country. Although the focuses of the schools were very different, there was the common interest in reforming curriculum to explicitly identify student growth goals and to work cooperatively with partner disciplines. Harvey Mudd, a highly selective small mathematics, science, and engineering college, was interested in developing an integrated core program that "balances teaching mathematics as mathematics with the need to establish more clearly the connections between mathematics and the applied sciences." Oklahoma State University was interested in developing a version of the West Point Core Curriculum model. Two liberal arts schools attended the Workshop/Retreat. Carroll College is developing an integrated two year program emphasizing breadth. They are using parts of COMAP's Principals & Practices of Mathematics text; University of Redlands focused on rearranging and integrating their Calculus II and III courses and also on various articulation issues.

[28]U.S. Military Academy, West Point, NY 10996

Carroll College Mathematics Curriculum Reform Project

John Scharf and Marie Vanisko[29]

1. Description of School and Student Audience

Carroll College is a small, Catholic, four-year institution offering professional programs in a strong liberal arts context. Located in Helena, the capital of Montana, Carroll draws sixty percent of its 1500 students from Montana and many of the rest from neighboring states in the northwest. In addition, Carroll has nearly 100 international students, representing 20 different countries. Approximately 80 students are in the mathematics major. About sixty percent of these are in a program that will lead to a professional engineering career, and many of the rest are in mathematics for secondary education. Nearly half of the graduates from our department pursue masters or doctoral degrees in engineering or mathematics, or they attend professional schools (e.g., medicine or law). At Carroll, the disciplines of mathematics, engineering, physics, and computer science are in one department. We see this as an advantage because faculty from a variety of disciplines work cooperatively together. Many of the ten faculty in this department have degrees in more than one discipline. They often teach courses from two or more areas to enrich our curriculum and course offerings.

Most of the students who enter Carroll to major in mathematics, engineering, or the sciences have strong high school preparation and require little or no remediation. Therefore, the curriculum that we design assumes that students have the mathematical maturity expected of first year students in a traditional college calculus course.

2. Current Status and Why Reform Is Needed

A primary goal for a curriculum in mathematics is to bring students to see the relationships among diverse mathematical topics and to appreciate their unifying themes. Without modern computer technology, this goal is difficult to realize because students are limited to problems that can be done by hand and because computer visualization as an aid to understanding is not available. Topics such as linear algebra and differential equations have to be postponed due to their computational complexity, resulting in a compartmentalization of mathematical topics in the curriculum. Problems are frequently limited to those that have closed form solutions and, as a result, applications have to be oversimplified. There is an emphasis on methods for finding closed form solutions at the expense of gaining understanding of concepts and principles.

Advances in computer technology are dramatically changing these circumstances. Problems that require more substantial manipulation can be done with the aid of the computer. Graphic visualizations are easy to generate as aids to understanding. More advanced topics can be introduced earlier in the curriculum because the computer can assist with computational complexities. Finally, problems can be addressed that do not need closed form solutions and, as a result, applications can be more interesting and realistic. Modern calculators and computers are changing the way we do mathematics and, consequently, they are impacting not only what we teach but also how we teach.

3. Model

The goals of our curriculum reform are:

1. to expose students to a wider range of mathematical topics in the first two years of the undergraduate curriculum,
2. to highlight themes that unify a variety of mathematical concepts,
3. to incorporate interesting and meaningful applications in every course taught,
4. to integrate the use of computer and information technology in a fundamental way throughout the curriculum, thus affecting what we teach, how we teach, and our expectations of the students,
5. to include explorations of ethical, social, cultural, and aesthetic issues associated with technological and scientific decisions,
6. to change teaching methodologies and course formats to include cooperative group learning activities and discovery-based laboratory exercises.

[29]Department of Mathematics, Engineering, Physics and Computer Science Carroll College Helena, MT 59625

The proposed curriculum for our first two years is laid out on the following page. It should be noted that the topics to be covered are integrated, not isolated.

4. Reception of Model

Our model is enthusiastically supported by the members of the Department of Mathematics, Engineering, Physics, and Computer Science and by the college administration. The administration sees our department as a model for other departments to follow in their curricular reform efforts.

Carroll College
Proposed Mathematics Curriculum
Years 1 and 2

30 APRIL 1995

TOPICS TO BE INTEGRATED	SEMESTER 1 6 cr. course	SEMESTER 2 6 cr. course	SEMESTER 3 6 cr. course	SEMESTER 4 6 cr. course
CALCULUS	Change, Difference/ Differential Equations	Accumulation of Change, Fundamental Theorem	Change, Vectors Functions of Vectors	Accumulation of Change for Vector Functions, Green's Theorem
LINEAR ALGEBRA	Matrices via Systems of Difference Equations	Gauss Elim., Inv., Det., Eigensysts., Transformations, Orthog./Proj.	Linearly Indep., Systems of DE's, Orthog./Proj.	Transformations
PROBABILITY and STATISTICS	Probability Markov Chains	Probability Distribution, Functions, Cummulative Distribution Functions, Moment Generating Functions, Correlation	Markov Chains, Curve-Fitting	Probability Distribution Functions, Cummulative Distribution Functions, Moment Generating Functions
GEOMETRY	Curve Analysis	Asymptotic Beh., Transf. Geom.	Surface Analysis	Transformational Geometry
SEQUENCES and SERIES	Finite Differences, Loc. Linearization	Geometric and Taylor Series	Loc. Linearization Generalized Taylor	
GRAPH THEORY	Adjacency Matrices, Digraphs	Adjacency Matrices, Digraphs	Networks, Applications	
NUMERICAL METHODS	Finite Diff's. Root Finding	Quadrature	Diff Eqns, Root Finding	Quadrature
COMPUTER SCIENCE	Spreadsheets, CAS, Syntax and Semantics, Graphics,Graphing Calculator, Internet	Control Structures, I/O, Matlab	Formalism, Additional Features, *Mathematica*	Packages, Capstone Computer Project
ENGINEERING DESIGN	Mini Design Problems	Iteration in Design Practice	Design Principles	Design Project
SOCIAL, ETHICAL, CULTURAL, AESTHETIC ISSUES	Research/Analysis of Issues, Decisions and Communication	Research/Analysis of Issues, Decisions and Communication	Research/Analysis of Issues, Decisions and Communication	Research/Analysis of Issues, Decisions and Communication

A Preliminary Plan for Curriculum Change at Harvey Mudd College: n-into-four

Robert Borrelli, Robert Keller, Michael Moody[30]

Harvey Mudd College is a small technical school and is part of the Claremont Colleges system. HMC has about 670 students; admission is very competitive–the average SAT score for the entering freshman in the fall of 1994 was 1410. About one-third of the incoming freshmen are National Merit Scholars. More than 40% of our graduates go on to obtain Ph.D.'s. Harvey Mudd College offers a B.S. degree in only six majors: Biology, Chemistry, Computer Science, Engineering, Mathematics, and Physics.

Math Backgrounds of HMC Freshmen. A year of high school calculus is a requirement for admission to HMC. Entering freshmen who have scored well on the AP Calculus BC are placed into either an enriched (Math 5) or regular (Math 4) multivariate course. The rest of the students are placed in a single variable course (Math 3), and go into Math 4 in the Spring.

Cooperation with other Departments. The six major departments at HMC have cordial relations, and discussions on curricular issues occur fairly often. There is also a Humanities/Social Sciences Department which enjoys the same cordial relations with the other departments. The College has recently appointed a Curriculum Planning Committee whose charge is to evaluate our current curriculum, moderate and lead the debate on curricular reform, and finally to propose concrete changes as necessary (large or small).

Facilities. HMC generally has very good support for science, mathematics, and engineering instruction. Apart from laboratories for the sciences and for research, there are several computing facilities, including two PC and Macintosh laboratories, a scientific computing laboratory with HP workstations, a computer science lab with Sun workstations, and a computer graphics lab with several Indigo-2 SGI computers. An ethernet network supports electronic communications, throughout the campus, including all dorm rooms. Several classrooms are equipped with computer projection systems.

The Core Curriculum at Harvey Mudd College

Students at Harvey Mudd College complete a common technical core curriculum, which consists of mathematics through linear algebra and differential equations, three semesters of physics (with laboratory), two semesters of chemistry (with laboratory), one semester of computer science, one semester of systems engineering, one semester of biology and two technical sophomore electives (Mathematics has as its sophomore elective Math 55: Discrete Mathematics).

Current Syllabi. In Appendix A are listed the current syllabi for the core mathematics and science courses required of all HMC students. These courses are Math 3, 4,(or 5), 73 and Math 82, Engineering 53, Computer Science 5 (or 6), and Physics 23, 24, 28 (lab), 51, and 53 (lab), Chemistry 21, 22, 25 (lab), 26 (lab), Biology 52, and 2 core electives.

Curricular Goals

The design of our curriculum will be shaped by a desire for a more creative synthesis between mathematics and applied fields of study. This synthesis should balance teaching mathematics as mathematics with the need to establish more clearly the connections between mathematics and the applied sciences. To accomplish this objective will require something more than merely exchanging mathematics for applications. Our revised curriculum should:

- Encourage the development of intuitive thinking
- Improve student's appreciation of mathematical rigor
- Decompartmentalize mathematics from its applied fields
- Improve our articulation, and in particular the timing of our introduction of mathematical topics to coordinate with other fields
- Develop a sense of how mathematics is used as a tool for modeling

[30]Harvey Mudd College, Claremont, CA 91711

- Develop a hands-on laboratory component for the introductory mathematics courses, combining physical experiments, computer analysis, and mathematical modeling
- Devote more time to group projects with a mathematical emphasis
- Regularly expose the student to open-ended problems

Goals for Student Growth. Apart from the specific content that we expect students to master, we hope that our revised curriculum will improve the development of students' mathematical reasoning skills: they should better understand mathematical precision, mathematical induction, proof, and logic. Students should also develop a better understanding of the process of constructing, analyzing, and interpreting mathematical models, and should be more competent in using computation as a tool for mathematical analysis and exploration. Another priority will be the improvement of students' ability to write mathematics–the clarity of mathematical expression through writing.

Topics in first two years. The following outline describes the topics that we feel should be included in the first four semesters.

- Elementary ODEs (very basic solution techniques; properties)
  - first order linear
  - separation of variables
  - existence and uniqueness
  - second order constant coefficients
- Complex numbers (series)
- Discrete dynamical systems
  - beginning block in common first-term course
  - connect with CS 5
  - biology (Bio 52)
  - Chemistry (patterns in reactions)
- Differential and integral calculus
  - limits
  - differentiation
  - integration rules
- Vector calculus
  - partial derivatives and chain rules
  - Taylor's Theorem
  - multiple integration
  - divergence, gradient, curl
  - line integral
  - vector theorems
- Linear Algebra
  - Systems of equations, determinants, matrices
  - Abstract linear algebra, vector space theory, operators
  - Spectral Theory of matrices
- Predicate logic
- Differential Equations
  - Systems of differential equations
  - Nonlinear differential equations
  - Dynamical systems (stability, bifurcation etc.)
- Probability and statistics
  - densities and distributions
  - expectation
  - Markov models
  - basic combinatorics
  - basic data analysis
  - linear regression
  - sample mean and standard deviation
  - transformations of data
  - independence
  - conditional probability
  - confidence intervals
  - least squares
- Numerical Methods

A Proposed Curriculum

How will we fit all of the above into four courses? If we are thinking about standard mathematics courses, in isolation, then it would be unworkable. We have designed our curriculum, therefore, on these assumptions: that we have a one-credit laboratory session between mathematics and the other departments. Major and minor topics will be divided between laboratory and lecture within the math courses, and some topics will be developed in non-mathematics courses, such as in CS 5 or E 53. For example, numerical methods, such as root-finding, quadrature, solution methods for ODEs and so on, will appear in the laboratory, or in computer science as programming exercises. Most of the topics from Probability and Statistics will be developed in the laboratories, or examples of applications in the problem exercises. The outlines that we give below will not show this level of detail.

Course A: Discrete Dynamical Systems and Single Variable Calculus
Lecture Topics
- Logic (2 weeks)
- Discrete DS (4 weeks)
- Calculus with simple ODEs (4 weeks)
- Integration (one and several dimensions) (4 weeks)

Course B: Vector Calculus
Lecture Topics
- Vectors, matrices, systems of equations (2 weeks)
- Vector calculus (5 weeks)
- Iterated integrals (2 weeks)
- Div, grad, curl (2 weeks)
- Line integral (1 week)
- Surface integral (1 week)
- Vector theorems (2 weeks)

Course C: Linear Algebra
Lecture Topics
- Linear Algebra
- Determinants
- Inverses
- Spectral theory
- Operators (Vector spaces)
- Function spaces
- Inner product space
- Orthogonality and projection

Course D: Differential Equations
Lecture Topics
- Modeling
- Constant coefficients
- Series solutions
- Systems
- Dynamical systems
- Stability/Bifurcation
- Periodic solutions
- Chaos

Other Directions

We are also exploring the possibility of developing a new course (or courses), which would combine ODE (M 82) and the sophomore systems Engineering (E 53). These two courses share much in common, and we feel that a synthesized course, taught by instructors from both departments, offers interesting and novel possibilities.

Appendix A: Current Technical Core Curriculum

MATHEMATICS

3. Calculus. Sets and logic, review of selected topics from single-variable calculus with emphasis on material not usually covered at the high school level, complex numbers, infinite sequences and series. Prerequisite: a year of calculus at the high school level. 4 credit hours. (First semester.)

4. Multivariable Calculus. Calculus of vector-valued functions, partial derivatives, multiple integrals, calculus of vector fields, introduction to probability. Prerequisite: Mathematics 3. 3 credit hours. (Students may not receive credit for both Mathematics 4 and Mathematics 5.) (Second semester.)

5. Multivariable Calculus. The material of Mathematics 4, together with logic and sets, limits of functions, continuity, intermediate and mean value theorems, and complex numbers. Prerequisite: Mathematics 3, or the equivalent. 4 credit hours. (First semester.)

73. Linear Algebra. Linear spaces, linear transformations and matrices, determinants, eigenvalues and eigenvectors, similarity and diagonal forms, and quadratic forms. Prerequisite: Mathematics 4 or 5. 3 credit hours. (Both semesters.)

82. Differential Equations. An introduction to the general theory and applications of differential equations; linear systems; non-linear systems and stability. Applications from engineering, physical science, and biological science. Prerequisite: Mathematics 73. 3 credit hours. (Both semesters.)

ENGINEERING

53. Introduction to System Engineering. An introduction to the concepts of modern engineering, emphasizing modeling, analysis, synthesis, and design. Applications to chemical, mechanical, and electrical systems. Prerequisites: sophomore standing and concurrent registration in Physics 51. 3 credit hours. (First semester.)

COMPUTER SCIENCE

5. Structured Programming and Problem Solving. Introduction to problem solving using the computer. Algorithms, data representation, and structuring. Use of programming languages and operating systems. Specification, testing, debugging, and documentation. Computing infrastructure at the College. 3 credit hours. (First semester).

6. Computer Problem Solving and Applications. An accelerated approach to computer problem solving, with emphasis on structured programming, style, and applications in engineering and the sciences. Use of operating systems and other software tools. Computing infrastructure at the College. 3 credit hours. (First semester.)

PHYSICS

23-24. Mechanics and Wave Motion. Kinematics, dynamics, linear and angular momentum, work and energy, harmonic and central force motion, waves and sound) and an introduction to special relativity. 2 credit hours. (First semester.) 3 credit hours. (Second Semester.)

28. Physics Laboratory. Experiments in mechanics using digital electronic measuring devices. Corequisite with Physics 24. 1 credit hour. (Second semester.)

51. Electromagnetic Theory and Optics. An introduction to electricity and magnetism leading to Maxwell's electromagnetic equations in differential and integral form. Selected topics in physical optics. Prerequisites: Physics 23-24 and Mathematics 3 and 4. 3 credit hours. (First Semester.)

53. Electricity and Optics Laboratory. Electrical and magnetic techniques in such measurements as the Hall effect and the earth's magnetic field. Introduction to electronics, including use of the oscilloscope and measurements on and RCL circuits. Experiments in physical optics, including studies of diffraction patterns. Prerequisite: Physics 51 or concurrently. 1 credit hour. (First semester.)

CHEMISTRY

21-22. General Chemistry. Stoichiometry, kinetic theory, phase behavior, equilibrium, bonding, thermodynamics, kinetics, and descriptive chemistry. 3 credit hours per semester.

25-26. Chemistry Laboratory. Laboratory taken concurrently with Chemistry 21-22. 1 credit hour per semester.

BIOLOGY

Biology 52. Introduction to Biology. Topics in the biology of molecules, cells, organisms, and populations, with some emphasis on the interfaces between biology and the physical sciences and engineering. Prerequisites: one semester of general chemistry and one semester of calculus. 3 credit hours. (Both semesters.)

Core Curriculum Reform Model for the Oklahoma State University

James Choike[31]

Description of School and Student Audience

Oklahoma State University (OSU) is a comprehensive land-grant university situated in a rural setting with an enrollment of approximately 18,000 students. The Department of Mathematics has the largest teaching staff (faculty, visiting faculty, graduate students, and part-time instructional staff) on campus and also teaches the greatest number of student credit hours of any department on campus. The Mathematics Department teaches 14% of all lower division credit hours and 8% of all undergraduate credit hours. More than three-quarters of all student credit hours taught by the department are lower division credits for non-majors in the colleges of Arts and Sciences, Engineering, and Business. The courses generating most of these student credit hours are:

- College Algebra, a 3-credit hour university general education requirement in mathematics for all students;
- Calculus I and II, the calculus sequence (each course is 5-credit hours) taken by mathematics majors and all science and engineering students;
- Differential Equations, a 3-credit hour course taken by mathematics, science, and engineering majors after completing Calculus II;
- Business Calculus, a 3-credit hour course in differential and integral calculus required of students majoring in College of Business degree programs.

Students entering the calculus sequence are, generally speaking, adequately prepared with Algebra I, Geometry, Algebra II, and Trigonometry taken at the high school level. Many students who anticipate majoring in mathematics, science, and engineering while still in high school, take a fourth-year course in calculus in high school. The majority of these students do not take AP credit for calculus, nor do they take a CLEP exam in calculus. They prefer, instead, to enroll in the OSU calculus sequence, starting with Calculus I.

Algebra I, Geometry, and Algebra II are required for admission to Oklahoma State University without deficiencies. Students may be admitted with one, possibly two curricular deficiencies, and they typically occur in mathematics. Intermediate Algebra (equivalent to high school Algebra II) is offered as a developmental/remedial three-hour course through our Extension College to enable students to remove an Algebra II deficiency. The grade students receive in Intermediate Algebra is counted like a three-credit course toward their grade point, but the course itself does not count as credit toward any degree on campus.

Description of Current Status and Why Reform is Needed

The following 5-course mathematics core is the academic path taken by mathematics, science, and electrical engineering majors:

- Calculus I, 5 credits
- Calculus II, 5 credits
- Differential Equations, 3 credits
- Linear Algebra, 3 credits
- Multivariate Calculus, 3 credits

All other engineering majors take "Engineering Statistics," a 3-credit hour course, taught by the Department of Statistics, in place of "Linear Algebra"; "Multivariate Calculus" is treated as an elective by the non-electrical engineering majors.

There is a general dissatisfaction among faculty with these courses and the dissatisfaction has grown more pronounced in recent years. A major issue centers around a belief that the current curriculum no longer meets or serves the needs of students who take these courses. Factors contributing to this general dissatisfaction center around questions faculty have about calculus reform, the proper use of technology in the classroom, the compartmentalized presentation of topics in and across these courses, and the changing role of service to client departments. These questions really boil down to the following:

[31]Department of Mathematics, Oklahoma State University, Stillwater, OK 74078

do these courses prepare students to use the mathematics that they encounter after they have completed the curriculum?

Recently, the College of Engineering removed "Linear Algebra" as a mathematics requirement for their non-electrical engineering majors and replaced it with "Engineering Statistics." This curricular change illustrates another dilemma imbedded in the current 5-course mathematics core. Linear Algebra is still seen as a major component of the mathematical training of all engineers. But, today, technology places additional demands on engineers to be able to collect and analyze data, and to be able to handle stochastic models. Consequently, course work in statistics has become increasingly important to engineering majors. An engineering degree plan, already packed with courses required by professional accreditation groups, is forced into an "either-or" decision when considering whether Linear Algebra or Statistics should be in the engineering mathematics core.

Another major problem with the current 5-course mathematics core is that multivariate calculus does not receive adequate coverage. Calculus II treats vector functions and elementary multivariate calculus, but, because this course also covers techniques of integration, the calculus of transcendental functions, sequences, and series, it sets a brisk and challenging pace for students. Calculus II closes with double and triple integrals, short of the important Green's and Stokes' theorems. Multivariate Calculus, a 3-credit hour course, is typically taken in the first semester of a student's junior year, more than a year and a half after completing Calculus II. This long time delay necessitates a significant review of topics previously seen in Calculus II. Although the treatment of these topics in Multivariate Calculus is in greater depth, once again, adequate coverage of Green's and Stokes' theorems does not quite materialize.

The Core Curriculum Model

Initial Considerations

Two critical questions should guide the development of a core curriculum in mathematics: what should be taught and how should it be taught? The identification of content and pedagogy threads in a core curriculum can be assisted by first identifying expectations in student academic development over time in the core. These expectations of academic development will be called "Student Growth Objectives." They are given below.

Student Growth Objectives: After completion of a core curriculum in collegiate mathematics, a student is expected to be able:

- to reason logically;
- to know what to do when they don't know what to do;
- to interpret and construct models;
- to do real-world problem-solving;
- to think with multiple representations;
- to communicate mathematically.

In addition, it is expected that students will

- see mathematics connected to other disciplines;
- learn in and with groups;
- increase in mathematical confidence;
- increase in positive attitudes about mathematics.

Content Curriculum Threads

Most discussions of content reform in mathematics usually get sidetracked into a discussion of topics; what should be covered and what should be left out? Most mathematicians, scientists, and engineers who take part in such discussions find that when all topics valued by the discussants are added to a list, the list is too large to cover adequately in the credit hours that are available. To a degree the reform calculus suffers from this affliction; namely, different people value different types of mathematics. The calculus reform was rooted in a call for a "lean and lively" calculus. Most reformed calculus texts have heeded the call for "lively," but many have missed on the call for "lean."

What should a student understand from a mathematics core curriculum? To avoid the entrapment of listing topics, one approach is to identify content curriculum threads that characterize mathematics. Mathematics is characterized by concepts, processes, and skills. "Concepts" are the basic and fundamental ideas of mathematics. "Processes" denote the ways in which mathematics is done. "Skills" refer to the procedures, rules, and algorithms that are a part of mathematics. All three of these levels are important and must be present in a student's understanding of mathematics. The OSU current core curriculum attempts to address the concepts of mathematics, but, frankly, it is overloaded on the skills and weak on the processes. The goal for reform is to design a core curriculum which presents a balance of concepts, processes, and skills that are tied to the student growth objectives listed above.

Starting with the three major levels of concepts, processes, and skills, there are many ways to build content threads for a mathematics core curriculum. A content thread is an on-going theme or strand, woven throughout the curriculum. Below is a framework for content threads to guide a reform of the 5-course core at OSU.

CONCEPT THREADS

- Representation (to interpret)
- Function: Discrete and Continuous, Linear and Non-linear
- Limit
- Probability
- Models (to interpret): Deterministic and Stochastic

PROCESS THREADS

- Reasoning
- Representation (to construct)
- Models (to construct)
- Communication

SKILL THREADS

- Algebra, includes linear algebra
- Differential and Integral Calculus
- Data Collection and Analysis
- Technology: Graphing Calculator, Computer Algebra Systems, and Spreadsheets

Some comments may help the reader to understand the above content threads under concepts, processes, and skills.

"Representation" denotes the languages of mathematics: numeric, symbolic, and graphic; also included under "representation" is writing and diagrams. "Representation," as a content thread, appears under both concepts and processes. As a concept, the goal is to teach students to interpret mathematics from the various ways of representing mathematics. The emphasis on "representation," as a process, is tied to the expectation of student growth to a higher level of being able to construct mathematics using various ways of representing mathematics.

"Function" is central to mathematics. The concept of function is first encountered by students in elementary algebra at the high school level when students are introduced to variables and dependency of variables. In collegiate mathematics function continues to be central, but now the emphasis shifts to deepening and refining the concept of function. At the collegiate level, the major threads of function are linear and non-linear functions, and discrete and continuous functions.

"Limit" is the fundamental concept of calculus, and analysis, in general. The fundamental applications of limit in calculus include the derivative and the integral. Notions of convergence and approximations cannot be made rigorous, or even intuitive, without an understanding of the concept of limit.

Students see examples of mathematical models at an early and elementary stage whenever they are presented with a function that represents or describes some phenomenon. "Models," as a content thread, refers to the notion of mathematical systems that represent or describe phenomena. At the concept level, the goal for models is to teach students to interpret and analyze models that are presented to them. "Models," similar to "representation," also appears at the process level. At this level the expectation on the student is greater, namely, to be able to construct models. There are two major types of models that the student must experience in the core curriculum: deterministic models and stochastic models.

"Reasoning," as a process thread, includes several types of reasoning: deductive reasoning (reasoning by rigorous proof), inductive reasoning (pattern recognition and the reasoning of investigation and exploration), and problem solving (the reasoning of algebra and heuristics).

"Communication," as a process thread, refers to all types of mathematics communication: communicating with writing and orally, both formally and informally, and communicating through working in and with groups.

The skill threads contain most of the topics one might find listed in the table of contents of a textbook. It has already been stated earlier in this paper that is not possible to cover all of these topics, especially using a compartmentalized approach at teaching a core curriculum. The topics to be presented must include the most important features of algebra, including linear algebra, differential and integral calculus, and the elements of data collection and data analysis. The content thread of technology at the skill level emphasizes the important role of technology in the core curriculum. Technology, however, is seen primarily as a tool in the learning of important mathematical concepts and the practice of important mathematical processes. Students should have opportunities to use graphing calculators, computer algebra systems, and spreadsheets as they learn and do mathematics in the core curriculum.

Curriculum Reform Goals

Since curriculum is the responsibility of the entire faculty, curriculum goals for reforming the current core curriculum at OSU must be endorsed by the departmental faculty. In the OSU Department of Mathematics, every faculty member takes the responsibility of maintaining the standards and the integrity of the curriculum very seriously. Goals for reform, by themselves, will not win the endorsement of the faculty. What is needed, additionally, is an accompanying implementation plan, which, inherent in its design, affords opportunities for appropriate faculty interaction, reflection, and involvement. Consequently, the goals listed below represent only one approach to reforming the OSU core curriculum. The next section details an implementation plan designed to provide opportunities for appropriate faculty involvement. It is expected that, through faculty involvement, the goals may evolve, and, in fact, even change, but, in the end, the faculty will have addressed the issues of needed reform for the benefit of student growth in mathematics.

GOALS

- To introduce modeling and projects throughout the mathematics core, and especially in the calculus sequence.
- To begin Calculus I with "Difference Equations".
- To implement a 3-into-2 curriculum transformation by transforming the fundamental content of "Differential Equations," "Linear Algebra," and "Multivariate Calculus" into two 3-credit hour courses.
- To increase OSU faculty awareness of issues of reform in mathematics content and mathematics teaching as a foundation for making needed curriculum reform at OSU.
- To strengthen connections with faculty in client disciplines.
- To identify and develop projects for students that cut across disciplines and courses with faculty in client or service departments.

An Implementation Plan

Oklahoma State University has a university-wide Honors Program. The Department of Mathematics participates in this Honors Program by offering selected courses for Honors credit. The courses that have been offered for Honors credit have been Calculus I and II, Differential Equations, and Linear Algebra. Class size in Honors sections are smaller, usually limited to approximately 25 students, and Honors sections are expected to be "honors" in the sense of offering students more academic challenge and enrichment. The Honors sections offer the opportunity to pilot test a curriculum based on the student growth objectives, the content threads, and the curriculum goals given above.

The implementation plan starts with a pilot run of a "new" Calculus I and II in an Honors section for this sequence. The pilot run will test the following:

- The use of a "reform calculus" text as only one of several resources for accomplishing curriculum goals. The Department of Mathematics has been involved with using and testing several of the currently available "reform calculus" texts. The general feeling from these trials is that, although there are merits to each text, no "reformed" text accomplishes exactly what the Department would like to see accomplished in its Calculus sequence. Generally speaking, that has always been true for most texts through the ages. However, the encyclopedic traditional texts had the virtue (or vice) of including everything any department might want to cover. A "reformed" text will be used accepting the realization that it will be supplemented when it falls short in areas of curriculum need.

- The use of the topic of "Difference Equations" to open Calculus I. The West Point 7-into-4 Core Curriculum takes this approach in their first course for cadets. There are content and pedagogical values for beginning with this topic. The topic is new for everyone in the class, including those who enter already knowing some calculus. As a result, this topic captures students' immediate attention and creates an even "playing field." Difference equations provides content connections to the concept of limit and, later, further connections when differential equations are introduced and discussed. With a treatment of difference equations, students have a valuable opportunity to compare discrete mathematics with continuous mathematics.

- The use of projects and modeling in Calculus I and Calculus II. To build the appropriate student growth, the OSU plan is to start with two projects in each course, with each project requiring about 3 days of class time. The first project or two may be no more than an extended word problem with the early goal of teaching students what a mathematical model is and how to interpret models. As the course progresses and students grow mathematically, later projects will require students to construct models.
- The use of technology, such as spreadsheets, graphing calculators, and computer algebra systems. Most students already have graphing calculators and own microcomputers with spreadsheet software. However, what has been lacking in the past, including the recent past, is the failure of the mathematics curriculum to capitalize on such technology. Part of the problem is that making use of such technology in a meaningful way requires planning time, something that is not always available to faculty. Efforts will begin to encourage faculty who find time to plan activities, exercises, or problems which use such technology to share their results with others.
- The use of gateways as a tool for reducing in-class time spent in needless review. A needless waste of valuable class time is to review topics that all students have already had. The West Point 7-into-4 Core Curriculum uses the tool of gateway exams to motivate students to do their review on their own outside-of-class time. A version of West Point's gateways will be implemented at OSU.
- To place an emphasis on student growth.
- To require more communication in writing and orally.
- To provide opportunities for students to work in and with groups. The use of projects will be a means of accomplishing student growth, more communication in writing and orally, and group work for students.

A critical feature of the pilot run of the Calculus sequence, as described above, will be the involvement of the Department's Curriculum Committee to document the progress of this test of a "new" curriculum for calculus. The documentation will detail all materials used, what worked, what didn't, why something worked or didn't, samples of student work, and student surveys of the "experiment." It is anticipated that the documentation, by itself, will not generate faculty endorsement of curriculum reform, but will provide a focal point and context for discussion of needed reform.

To increase OSU faculty awareness of issues of reform in mathematics content and mathematics teaching several specific actions will be implemented. First, the pilot run of Calculus I and II together with a documentation of results is one step in increasing faculty awareness of needed curricular reform. Other steps that are being planned include:

- a regularly scheduled Mathematics Issues Seminar with specific topics to attract a large body of the faculty. Examples of topics are: progress reports on the pilot run in the Honors section of the Calculus sequence; West Point 7-into-4 Core Curriculum, what is it and why; what are your content threads of the Mathematics Core; have faculty from client departments describe problems they would like to see discussed in the Core; etc.
- an E-mail home page. A home page set up, by mathematics course, which provides an opportunity for faculty to easily share ideas and tips about improving the teaching and student growth in the given course.
- a Faculty Retreat on an Issue of Curriculum Reform. Set aside a large block of time for discussion of a curriculum reform issue. The focus of the discussion could begin with an expert invited to be a speaker as part of the Department's Colloquium Series.

Plan to Integrate Calculus II and III Curricula at University of Redlands

J. Beery, P. Cornell, A. Killpatrick, and A. Koonce[32]

Description of School and Student Audience

The University of Redlands is an independent, coeducational, comprehensive university located in the city of Redlands, California. It enrolls 1500 undergraduates in the arts and sciences and in small professional programs in communicative disorders, business administration, and music. The student/faculty ratio is approximately 13:1. The University operates under a 4-1-4 academic calendar, with 13-week fall and spring semesters and a 3 1/2-week January term.

The Department of Mathematics has eight full-time faculty members, including two professors, three associate professors, two assistant professors, and one lecturer. An additional mathematics professor serves as the University's Director of Academic Computing. The Department offers two degrees, a B.S. degree in mathematics and a B.S. degree in mathematics leading to a California secondary teaching credential. The Department graduates approximately nine majors per year, with approximately one third of these pursuing the secondary teaching credential. The Department also graduates approximately 13 mathematics minors each year.

In addition to offering courses supporting its major and minor programs, the Mathematics Department offers service courses for biology, chemistry, physics, environmental studies, computer science, economics, business administration, and liberal studies (elementary education) majors, as well as courses satisfying the University's general education requirement in Quantitative Reasoning. In order to fulfill the Quantitative Reasoning requirement, most students take a Finite Mathematics course which has a minimal high school algebra prerequisite. Approximately 200 students per year, however, take at least one calculus course. We offer six sections of Calculus I with approximately 24 students per section, five sections of Calculus II with approximately 22 students per section, and three sections of Calculus III, with approximately 12 students per section, each year. We also offer a PreCalculus course for students who wish to take Calculus I but are not adequately prepared for it. Our calculus sequence serves as our introductory mathematics sequence, and we attempt to convince as many students as possible to study calculus and to entice as many calculus students as possible to major or minor in mathematics.

There is just one calculus track, and, as a result, the students in the first year calculus courses (Calculus I and II) have varied interests and abilities. Although many intend to major in mathematics, science, or computer science, a fair number of these students plan to major in business or in one of the social sciences. A large number are preparing for graduate study in medicine, dentistry, or other health professions. Approximately one-third of the students in Calculus I have studied calculus previously. With the exception of new students who place into Calculus III, we find our students' high school preparation to be generally weak; students often need extensive review of the pre-calculus and calculus topics they studied in high school. The students who continue from Calculus II to Calculus III are primarily mathematics, physics, or chemistry majors or minors.

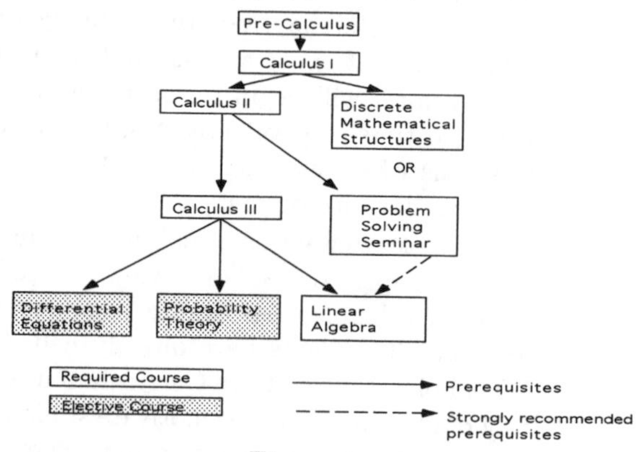

Figure 1

Description of Current Status and Why Reform Is Needed

Our core curriculum for the first two years cur-

[32]Department of Mathematics, University of Redlands, Redlands, CA 92373

rently consists of Calculus I, II, and III, Problem Solving Seminar (or Discrete Mathematical Structures), and Linear Algebra. (See Figure 1.) We discuss the latter three courses first.

Problem-Solving Seminar. The January term Problem-Solving Seminar is our "bridge to abstraction" course, emphasizing proof techniques as well as problem-solving strategies. Students learn these skills "in context"; an appropriate topic, such as graph theory or knot theory, is selected each year based on student and faculty interest. The course enrolls approximately 26 students per year with nearly all going on to major or minor in mathematics. As shown in Figure 1, the prerequisite for the Problem-Solving Seminar is Calculus II. In most years, this prerequisite is meant only to ensure a certain level of mathematical maturity; although the majority of Problem-Solving Seminar students have just completed Calculus III, we sometimes encourage talented Calculus I students to take the course during the January term immediately following their Calculus I course.

Discrete Mathematical Structures. Beginning in Fall, 1995, we will offer a lower division Discrete Mathematical Structures course for students interested in mathematics and computer science. Students with some calculus experience may take this course before beginning the calculus sequence, simultaneously with a calculus course, or as a "break" from the calculus sequence. Mathematics majors and minors may take Discrete Mathematical Structures instead of the Problem Solving Seminar.

Linear Algebra. Most students proceed from Calculus Ill to Linear Algebra and/or Differential Equations (spring semester courses) or to Probability Theory (fall semester course). Since Linear Algebra is a prerequisite for many upper division mathematics courses and since it appeals to so many potential mathematics majors, we strongly encourage students to take it as early as possible. In fact, the curriculum reform plan we introduce here allows students to study linear algebra topics and to take the Linear Algebra course even earlier than they previously could.

Our sophomore-level Linear Algebra course enrolls approximately 26 students per year, and, in combination with the Problem-Solving Seminar (or the Discrete Mathematical Structures course), is intended to prepare students for abstract thinking and theorem-proving in their upper division mathematics courses. The course covers the standard elementary linear algebra topics, but also emphasizes applications and uses computing as an integral tool. The course meets in a classroom in which each two students share a computer, enabling students to use the MATLAB package virtually every class day for computations, for formulating and testing conjectures, and for exploring applications.

In fact, all sections of our linear algebra, differential equations, calculus, and pre-calculus courses meet in classrooms equipped with one computer for each two students. We currently have two such classrooms, set up with University and NSF funding. These classrooms also are used for afternoon and evening tutorial sessions.

Calculus sequence. The three-semester calculus sequence serves as the backbone of our core mathematics curriculum. This is due primarily to the high demand for calculus instruction from partner disciplines, and our reluctance to track entering students by discipline or ability in different mathematics sequences or even in different calculus sequences. The most practical reason for not offering various mathematics or calculus tracks is that, with only 150 or so first year students (and 200 students in all) in calculus courses each year, offering only a few sections each of various types and levels of mathematics courses would create severe scheduling problems for students. Our main reason for not tracking students, however, is that we want students to have as much flexibility as possible in choosing a major—that is, we do not want entering students to have to decide on the first day of college if they are mathematics or physics or economics majors and we want students to be able to switch majors as easily as possible during their first two years of college. Since there is a large demand for calculus instruction from client disciplines and since calculus is an interesting, vital, and useful area of mathematics, we intend to retain the calculus sequence as our introductory mathematics sequence. The main disadvantages of having a single introductory mathematics sequence emphasizing calculus are that students may not get an accurate view of the nature and scope of mathematics and some students may not have the opportunity to study the type of mathematics that appeals most to them. The Problem-Solving Seminar and Discrete Mathematical Structures courses were created in part to address these concerns. The present curriculum plan attempts to address these issues not only by making our calculus courses more appealing, but also by allowing students to study linear algebra topics earlier than they previously could.

During the academic years 1992–93 and 1993–94, we used the innovative, computer-based Calculus in Context curriculum [6] in all of our Calculus I and II courses. Created by members of the mathematics departments of the Five College Consortium (Amherst, Hampshire, Mount Holyoke, and Smith Colleges, and the University of Massachusetts, Amherst) with funding from the National Science Foundation (NSF), this curriculum develops calculus concepts in the context of scientific applications, and uses computer programs and graphing packages as exploratory tools. For example, on the first day of Calculus I class, students begin to construct a model for a measles epidemic. After setting up rate (differential) equations for the susceptible, infected, and recovered populations, the students use these equations to predict future sizes of the populations, first by hand and then using simple True BASIC computer programs. As they use the computer to calculate and plot future population sizes, they soon notice that their approximations seem to approach limiting values when they recompute them using smaller and smaller step sizes. In this way, students discover, and gain an intuitive understanding of, the limit process. Later in the first course, the derivative is introduced as the slope of the straight line the students see when they look at a small section of the graph of a (locally linear) function under a "computer microscope." It became apparent early on that our students were gaining a better understanding of the limit process and of the derivative as slope than they ever had in the past. Indeed, the Calculus in Context curriculum forces students to focus on concepts and to communicate clearly—both orally and in writing—about them. As an added bonus, the percentage of students in fall semester Calculus I courses who continue to spring semester Calculus II courses has increased from 45% to 85% since we instituted the Calculus in Context curriculum.

We have been quite pleased with the Calculus in Context curriculum in our Calculus I course and intend to keep it in place. However, although the percentage of students in Calculus I who continue to Calculus II has increased dramatically since we instituted the Calculus in Context curriculum, the number of students who continue from Calculus II to Calculus III has not increased as substantially. Currently, nearly 80% of our Calculus I students proceed to Calculus II, whereas only about one-third of our Calculus II students go on to Calculus III. We believe that this drop off in retention is due in large part to our continuing inability to make Calculus II an interesting and coherent course in which students can be successful. Both the traditional Calculus II syllabus and the more innovative Calculus in Context syllabus for the course (which includes dynamical systems in addition to single variable integration and Taylor series approximations) seem to students to be collections of unrelated topics, which become successively more difficult and which culminate in a topic most Calculus II students find nearly impossible to understand: Taylor series approximations.

In order to address students' concerns about the Calculus II course, we have tried changes in both pedagogy and content. Prior to adoption of the Calculus in Context curriculum, we used computer demonstrations as well as a series of laboratory sessions, some computer-based, in our Calculus II courses. (See [1], [2], [4], and [8] for details.) While this seemed to improve students' attitudes, understanding, and communication skills, the retention rate did not increase significantly. The Calculus in Context Calculus II curriculum, which we began using in Spring, 1993, includes exciting topics which build on students' work with differential equations in Calculus I. Unfortunately, during the three semesters in which we used it, we found that it greatly overestimated our students' prerequisite knowledge, making the Calculus II course as disjointed and difficult for them as our previous versions of the class. Furthermore, students who had studied Calculus I in high school or at another college or university were at an even greater disadvantage in this course, because of their lack of experience with differential equations. These problems precipitated what we hoped would be a temporary retreat to a more traditional Calculus II course.

During the academic year 1994-95, we used the Calculus in Context curriculum in the Calculus I course, but, in order to make it possible to place entering students with a traditional Calculus I background in Calculus II and to extend "reformed" calculus instruction to our Calculus III course, we used the more conventional NSF-funded Calculus Consortium at Harvard curriculum for the Calculus II and Ill courses [3]. In addition, in our PreCalculus course, we used preliminary materials written by members of the Bridge Calculus Consortium at Harvard, along with the laboratory manual, *Pre-*

calculus in Context: Functioning in the Real World [5]. Although our experience with the Calculus in Context curriculum and, more generally, with technology in the classroom led us to adapt the Harvard materials in a manner which emphasizes discovery-based learning and in which computing is used as an integral tool, we have not been satisfied with the Harvard curriculum's traditional approach to single-variable calculus. Our only misgiving about our Calculus III course, which has traditional multivariable calculus content and in which we've used the *Maple* computer algebra system for two years and the Harvard Consortium's preliminary *Multivariable Calculus* text for one year, is that we try to cover too much material in too little time, resulting in inadequate student understanding. The present curriculum reform plan attempts to remedy this situation by distributing multivariable calculus topics throughout the Calculus II and III curricula.

Our immediate goal then is to make our Calculus II course more intrinsically interesting, coherent, and accessible than it has been, with the ultimate goal of enticing more Calculus II students to continue their study of mathematics. This means, of course, that we must ensure that the core curriculum courses that follow Calculus II, especially Calculus III, also are attractive courses for students. Hence, while our main objective for our core curriculum is to provide students with a solid foundation in mathematical ideas, methods, and thought processes within a reasonable time frame, we also want to help ensure that students continue their mathematical study by making the core mathematics curriculum, and especially our Calculus II course, as appealing and as flexible as possible for them.

Model

Goals for student growth

Members of the University of Redlands mathematics faculty are concerned primarily that our students understand mathematical concepts and that they learn how to think mathematically—that is, analytically, creatively, logically, abstractly, and, when appropriate, concretely. Our goal is for students to develop mathematics skills and, more generally, thinking skills they actually can use both in and beyond our classrooms. We want our students to become independent learners and thinkers who take responsibility for their own learning, who explore and discover mathematics on their own, and who take risks in problem solving. Such students must be able to learn in a variety of contexts, including abstract, concrete, and applied settings, and by various methods, including exploring examples, solving problems, reading, discussing and listening. Furthermore, we believe that expecting students to communicate mathematics clearly—both orally and in writing—not only deepens their understanding of mathematical concepts, but builds their confidence and gives them practical skills which they will use both in and after college. Most importantly, we want our mathematics students to develop enough ability, confidence, and enthusiasm not only to continue to the next mathematics course, whatever it may be, but to continue to learn and use mathematics throughout their lives.

We, along with many other mathematics educators across the nation, have come to believe that one of the best ways to ensure that individual students grapple with, understand, and learn to use mathematical ideas is to facilitate for them what our colleagues in education call constructive learning, or what we in mathematics tend to call discovery-based learning. This view, which is based on extensive study of existing programs as well as on our own experiences in incorporating computer use into our calculus, differential equations, linear algebra, and other courses, is consistent with the recommendations made in such publications as *Moving Beyond Myths: Revitalizing Undergraduate Mathematics* [7]. In recent years, we have attempted to facilitate discovery-based learning in all of our courses, but especially in our core curriculum courses, primarily through computer activities and other cooperative learning strategies.

Curriculum Reform Plan

While we believe that our recent improvements in pedagogy are helping us achieve our general goals for student growth as well as our more specific goal of making our core curriculum more effective and attractive for students, we feel that changes in content also are necessary in order to achieve this latter goal. As described above, we believe that the weak link in our core curriculum is the Calculus II course. In order to make this course more attractive, effective, and rewarding for students than it currently is, we propose the following integration of our Calculus II and III curricula.

Calculus II:

1. Single and multiple integration

Integration as accumulation, introduced in the context of applications
Fundamental Theorem of Calculus
Techniques of integration: substitution, parts
Brief introduction to multivariable functions and partial differentiation
Multiple integration
Nonrectangular coordinate systems

2. Introduction to vectors

Dot product and projections, cross product
Equations of lines and planes in three-dimensional space

Calculus III

1. Vector calculus

Multivariable functions (again) and their limits
Partial differentiation (again), and the chain rule
Gradient, directional derivative
Parametric equations
Line and surface integrals
Stokes', Divergence, and Green's theorems

2. Sequences and series

Sequences, series, power series, Taylor series
Transcendental functions

Both courses would continue to be computer-based, employing True BASIC computer programs and the *Maple* computer algebra system. While the actual topics proposed for the new Calculus II and III courses are quite traditional, we would continue to emphasize discovery-based learning in which computing is used as an integral tool, and, whenever possible, we would continue to introduce concepts in the context of scientific applications. Themes carried over from Calculus I to Calculus II and III would include successive approximation and multivariable functions, but less attention would be paid to differential equations in the new Calculus II and III courses. This should make it possible for students to succeed in the new Calculus II course whether their previous course was our Calculus in Context Calculus I course or a traditional high school or college Calculus I course.

ADVANTAGES OF CURRICULUM REFORM PLAN

We believe that the topics we've included in the Calculus II course form a more coherent whole than do the topics in our current Calculus II curriculum. We also believe that we have selected interesting, accessible concepts for the Calculus II course. The Calculus III topics also are interesting, of course, but we have found that they are more difficult for students than those we've included in the Calculus II course. We anticipate, however, that by the time students reach the Calculus III course, they will be able to understand these concepts. In particular, we predict that students will have a much better chance of understanding Taylor series approximations if their introduction is delayed to Calculus III. We also anticipate that students will understand the traditional multivariable calculus topics better if they are distributed throughout the Calculus II and III curricula. In summary, we believe that the proposed Calculus II and III curricula are more commensurate with students' interests and abilities than are our current curricula for these courses, and we expect students to be more successful in our new versions of these courses.

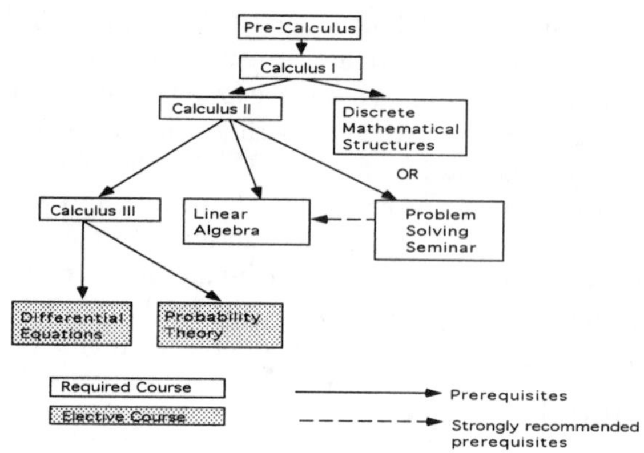

Figure 2

Our curriculum plan also has the advantage of allowing students to progress straight from Calculus II to Linear Algebra, as shown in Figure 2. (We note also that introducing vector topics at the end of Calculus II will give us an opportunity to introduce matrices and determinants in that class.) Of course, students also may proceed to Calculus III or to both Linear Algebra and Calculus III. This makes their core curriculum schedule slightly more flexible. We hope that reaching Linear Algebra sooner, together with the option of enrolling in our new Discrete Mathematical Structures course early on, ensures that more students who prefer discrete mathematics and algebra over calculus continue to study mathematics.

Note also that this curriculum plan ensures that physics students get some of the calculus concepts they need—most notably vectors and multiple integration—earlier than they previously did.

DISADVANTAGES OF CURRICULUM REFORM PLAN

The primary disadvantages of the plan are that 1) new students who formerly placed into Calculus III now must begin with Calculus II, 2) increased flexibility in student schedules may make scheduling courses more difficult for us, and 3) we have been unable to find textbooks that support this sequence of topics.

Virtually all students who place into Calculus III have Advanced Placement (AP) Examination scores of 4 or 5, and are awarded four to eight units of credit at the University of Redlands. We hope that these credits and perhaps the waiver of an elective in the mathematics major or minor would be enough to entice them to major or minor in mathematics. Nevertheless, we view 1) as a serious disadvantage of our plan.

We are certain we can deal with 2), however, as of this writing, we have been unsuccessful in overcoming 3).

RECEPTION OF MODEL BY DEPARTMENT, PARTNER DISCIPLINES, ADMINISTRATION

The other members of the Department of Mathematics have approved our plan to revise the content of Calculus II and III, but share our concerns about choosing textbooks for these courses, placing new students in these courses, and scheduling mathematics courses for optimal student flexibility.

Faculty in partner disciplines are largely responsible for the continuing centrality of the calculus in our introductory mathematics sequence. The majority of these faculty have supported our move toward an applications and computer-based calculus curriculum. In fact, many of them participated in a two-day workshop designed specifically to introduce faculty in client disciplines to the Calculus in Context curriculum and facilitated by the mathematics faculty during Summer, 1993. Physics faculty should appreciate students' earlier introduction to multiple integration and to vectors, which would occur now in Calculus II rather than in Calculus III.

The University's administration has been supportive of the mathematics faculty's curricular plans, and especially of our involvement in the calculus reform movement.

Contact Person

Dr. Janet L. Beery
Department of Mathematics
University of Redlands
1200 E. Colton Ave.
Redlands, CA 92373
E-mail: beery@ultrix.uor.edu
FAX: (909)793-2029
Office phone: (909)793-2121, extension 3118

Bibliography

[1] Beery, Janet L. Calculus with Weekly Exploratory Laboratories. *PRIMUS.* June, 1993.

[2] Beery, Janet L. My Second Favorite Lab–To This Point: Discovering Integral Formulas. *Computer Algebra Systems in Education (CASE) Newsletter.* July, 1993.

[3] Calculus Consortium based at Harvard. *Calculus.* 1994. Multivariable Calculus (Preliminary Version). 1995. New York: John Wiley & Sons.

[4] Cornez, Richard, Janet Beery, and Mary Scherer. A Computer-Based Calculus Curriculum. *College Teaching.* Spring, 1993.

[5] Davis, Marsha J., Judy Flagg Moran, and Mary E. Murphy. *Precalculus in Context: Functioning in the Real World.* 1993. Boston: PWS Publishing.

[6] Five College Calculus Project. *Calculus in Context.* 1995. New York: W.H. Freeman.

[7] *Moving Beyond Myths: Revitalizing Undergraduate Mathematics.* 1991. Washington, DC: National Academy Press.

[8] Scherer, Mary, Janet Beery, and Richard Cornez. Starting Small: Gradual Introduction of Computers into Calculus Courses. *PRIMUS*. June, 1993.